Research Analytics

Boosting University Productivity and
Competitiveness through Scientometrics

T0313101

Data Analytics Applications

Series Editor: Jay Liebowitz

PUBLISHED

Actionable Intelligence for Healthcare
by Jay Liebowitz, Amanda Dawson
ISBN: 978-1-4987-6665-4

Big Data Analytics in Cybersecurity
by Onur Savas and Julia Deng
ISBN: 978-1-4987-7212-9

Big Data and Analytics Applications in Government: Current Practices and Future Opportunities
by Gregory Richards
ISBN: 978-1-4987-6434-6

Data Analytics Applications in Education
by Jan Vanthienen and Kristoff De Witte
ISBN: 978-1-4987-6927-3

Data Analytics Applications in Latin America and Emerging Economies
by Eduardo Rodriguez
ISBN: 978-1-4987-6276-2

Intuition, Trust, and Analytics
by Jay Liebowitz, Joanna Paliszkiewicz, and Jerzy Gołuchowski
ISBN: 978-1-138-71912-5

Research Analytics: Boosting University Productivity and Competitiveness through Scientometrics
by Francisco J. Cantú-Ortiz
ISBN: 978-1-4987-6126-0

Sport Business Analytics: Using Data to Increase Revenue and Improve Operational Efficiency
by C. Keith Harrison, Scott Bukstein
ISBN: 978-1-4987-8542-6

Research Analytics

Boosting University Productivity and Competitiveness through Scientometrics

Edited by
Francisco J. Cantú-Ortiz

CRC Press
Taylor & Francis Group
Boca Raton London New York

CRC Press is an imprint of the
Taylor & Francis Group, an **informa** business

AN AUERBACH BOOK

CRC Press
Taylor & Francis Group
6000 Broken Sound Parkway NW, Suite 300
Boca Raton, FL 33487-2742

First issued in paperback 2022

© 2018 by Taylor & Francis Group, LLC
CRC Press is an imprint of Taylor & Francis Group, an Informa business

No claim to original U.S. Government works

ISBN 13: 978-1-03-247652-0 (pbk)
ISBN 13: 978-1-4987-8542-6 (hbk)

DOI: 10.1201/9781315155890

Library of Congress Cataloging-in-Publication Data

Names: Cantú-Ortiz, Francisco J., 1952- editor.
Title: Research analytics : boosting university productivity and competitiveness through scientometrics / edited by Francisco J. Cantú-Ortiz.
Description: Boca Raton, FL : Taylor & Francis, 2018. | Includes bibliographical references.
Identifiers: LCCN 2017019860 | ISBN 9781498785426 (hb : alk. paper)
Subjects: LCSH: Science indicators. | Big data. | Universities and colleges.| Bibliometrics.
Classification: LCC Q172.5.S34 R43 2017 | DDC 001.4/33--dc23
LC record available at https://lccn.loc.gov/2017019860

Visit the Taylor & Francis Web site at
http://www.taylorandfrancis.com

and the CRC Press Web site at
http://www.crcpress.com

To the lovely memory of Dr. Eugene Garfield (1925–2017).
For his legacy and foundational contributions to the field of Scientometrics.

I would like to thank my wife Carmen and our children Francisco, Héctor, Eduardo, and Marycarmen. Their prayers and love made it possible to work things out. My gratitude goes to Joseph for his company and guidance in times of joy, hope, and dejection.

Contents

SECTION II APPLICATION OF SCIENTOMETRICS TO UNIVERSITY COMPETITIVENESS AND WORLD-CLASS UNIVERSITIES

Foreword

"Whatever you can do, or dream you can, begin it. Boldness has genius, power and magic in it." These are the words of Johan Wolfgang von Goethe on the subject of commitment to an idea. He also notes that once one has embarked on the process of fulfillment of a dream or the realization of an idea, "All sort of things occur to help one that would never otherwise have occurred. A whole stream of events issues from the decision, raising in one's favor all manner of unforeseen incidents and meetings and material assistance, which no man could have dreamed would have come his way." For those of us who worked directly under Eugene Garfield, the man whose bold foray into the world of citation indexing laid the groundwork for all subsequent Bibliometric and Scientometric analyses and studies, these words have special meaning.

Dr. Garfield may have envisioned the digital incarnation of the progenitor of all bibliographic databases, the *Science Citation Index*, but even his foresight would only have glimpsed, as through a glass darkly, the present world of Bibliometric and Scientometric analysis we inhabit.

Each of the resources under discussion in this volume, including the mighty search engine Google, can trace its lineage back to Garfield's seminal work. It was his dream to create an index of ideas taking into account fully and completely all the associations between and among each of these ideas. Finding highly relevant studies quickly by searching the meticulously indexed and curated cited references from each source item was the beginning. Using these data as the fundamental basis for Bibliometric and Scientometric studies and analyses was the next step. In the early days, large-scale studies were cumbersome affairs employing multiple research associates and huge commitments of computer resources. Today, however, anyone with access to the Internet can survey millions of records and produce what pass for credible results.

Bibliometric and Scientometric results that actually *are* credible are another matter entirely, and the need for credibility has never been higher. Reputations, funding, careers, and national research initiatives pivot on the basis of Bibliometric and Scientometric data. Indeed, the credibility of scholarly research itself may rely on these data.

The vast enterprise of scholarly research and intellectual exploration has at its core the integrity of the source material on which it is based and from which it draws inspiration for further work. Analyses of these sources that lead to conclusions affecting all of humanity and the living systems on which humanity depends must proceed with extreme care and with the utmost integrity. Everything, however, depends on the credibility of the source(s) of the data under consideration.

Eugene Garfield knew this. Those of us who worked with him in those early days, long before the Internet was the medium of the masses, are intimately familiar with the essential need for accuracy, credibility, and, most importantly, stability in the datasets with which we work.

Dr. Garfield's death on February 26, 2017, marks the end of an era, one in which the world of scholarly inquiry and analysis underwent enormous change. The potential for discovery and analysis locked within the pages of the *Science Citation Index* has now been released into the ether for all to consider. Our mission, especially considering the ideals of the father of Bibliometrics and Scientometrics, is clear. First, we must base our analyses on the most credible sources, sources that have been indexed and curated with great care and integrity. Second, we must strive for utter clarity in the results that we derive from these monumental resources. This is how Garfield approached his work: boldly and with the strongest intent. It is our duty to follow his example.

James Testa
VP Emeritus
Clarivate Analytics

Preface

It was December 12, 2015, when I first learned about the Data Analytics Applications book series initiative when I visited Professor Jay Liebowitz, book series editor, at his home in Rockville, Maryland. When Liebowitz described the applications that he had in mind, I suggested that Data Analytics and Scientometrics could be a theme of interest for the book series. Liebowitz was very supportive of the idea, which finally took form after consultations with CRC Press editorial committee in New York. Data Analytics has become a hot topic in recent years attracting attention from academia, business, and industry. It comprises a set of techniques, methods, and applications from disciplines such as artificial intelligence, machine learning, statistics, pattern recognition, computer science, database systems, Big Data, and other fields and is enabled by dramatic progress on storage capacity and processing speed of digital technology. Data Analytics has reached a level of maturity whose methods and techniques have made it possible to treat and handle huge amounts of data in order to find useful patterns of knowledge and information that are relevant for decision-making and problem solving in academia, business, and industry.

We call the application of Data Analytics to Science *Research Analytics*. The phenomenon of data growth is also present in Science, as the number of scientific papers published doubles every 9–15 years, and the need for methods and tools to understand what is reported in scientific literature becomes evident. Research Analytics utilizes data science, Bibliometric databases, Scientometric methods, and research intelligence to discover and learn patterns of knowledge and trends in scientific progress and to identify researchers and institutions responsible for those advances. One application of Research Analytics for the last two decades has been the publication of big league tables of universities and knowledge institutions based on research indicators complemented with academic criteria about university demographics such as faculty size, student enrollment, international outlook, and industry collaboration, to mention a few. This is also complemented with alternate criteria that come for instance from traffic on the web or sustainability engagement of institutions.

An objective of this book is to assist students, professors, university managers, government, industry, and stakeholders in general, understand which are the main Bibliometric databases, what are the key research indicators, and who are the main players in university rankings and the methodologies and approaches that they employ in producing the tables.

Finally, this book intends to pay a humble tribute and homage to the memory of Dr. Eugene Garfield, founding father of the field of Scientometrics.

I hope you will enjoy reading this book and find it useful in your academic or professional responsibilities.

Francisco J. Cantú-Ortiz
Monterrey, Mexico

Acknowledgments

First, I would like to thank Professor Dr. Jay Liebowitz, Distinguished Chair of Applied Business and Finance, Harrisburg University of Science and Technology, and CRC Press book series editor in Data Analytics Applications, for the kind invitation to edit the present book on Research Analytics. His mentoring, advice, and friendship over the past decades have illuminated my academic career and taught me to understand and enjoy university life through his sensitiveness in human relations and his international leadership, knowledge, and expertise in knowledge management and related fields and especially for showing me the importance of a sense of humor and smile in everyday life. I also thank his wife Janet and his sons, parents, and in-laws. I would like to express my gratitude to our publisher Taylor & Francis/CRC Press, especially John Wyzalek, Senior Acquisitions Editor; Linda Leggio, Project Editor; and Adel Rosario, Project Manager at Manila Typesetting Company; and for their professional advice and guidance in editing the book and having it ready for publication.

A special thank you goes to each of the chapter authors, whose knowledge and well-known international leadership in the topics that they write about make the core and substance of this book. Their commitment, expertise, and dedication were fundamental in having this book take its final form. Any errors remain mine.

I appreciate the support from my university Tecnológico de Monterrey, and especially Salvador Alva Gómez, Miguel Schwarz Marx, David Noel Ramírez, David Garza, Arturo Molina, and Víctor Ortiz, for the backup I received from them in preparing this book. I also thank my team at the Research Office at Tecnológico de Monterrey for their valuable assistance in elaborating and revising the materials, principally James Fangmeyer Jr., Héctor G. Ceballos, Nathalíe Galeano, Érika Juárez, Yebel Durón, Liliana Guerra, and Bárbara Lancho.

Finally, I would like to acknowledge all of you who read the book. Your comments and suggestions will be highly appreciated.

Editor

Professor Francisco J. Cantú-Ortiz holds a PhD in artificial intelligence from the University of Edinburgh, United Kingdom; an MSc in computer science from North Dakota State University, United States; and a BSc in computer systems engineering from Tecnológico de Monterrey (ITESM), Mexico. He is a Full Professor of Computer Science and Artificial Intelligence and Associate Vice-Provost for Research and is responsible for World University Rankings at ITESM. He has published various scientific articles in international journals and conferences, and edited diverse books and conference proceedings in artificial intelligence, knowledge-based systems, and research management. He has served as President of the Mexican Society for Artificial Intelligence; Conference Chair of various artificial intelligence and expert systems conferences, including the Mexican Conference on Artificial Intelligence; and Local Chair of the International Joint Conference on Artificial Intelligence 2003. He is accredited a National Researcher (SNI) by the National Council for Science and Technology (CONACYT) in Mexico. His research interests include knowledge-based systems, machine learning, data mining, and data science applied to research intelligence and science and technology management. He has an interest in epistemology and the philosophy of science and religion (http://semtech.mty.itesm.mx/fcantu/).

Contributors

Juan M. Ayllón is a Research Fellow and a PhD candidate in the field of Bibliometrics and scientific communication at the Universidad de Granada (UGR) (Predoctoral Research Grant BES-2012-054980 granted by the Spanish Ministry of Economy and Competitiveness). His earlier degrees in library and information science are from the same university. He is also a member of the EC3 Research Group.

Atilio Bustos-González holds a bachelor in bibliotechnology, from the University of Chile; a master in management and university policy from Polytechnic University of Catalonia, Spain; and a doctorate in information sciences (Scientometry) from the University of Extremadura, Spain. He is an Associate Researcher with the Scimago Research Group. He was a member of Scopus' Content Selection Committee from its foundation in 2005 until 2012. He has been the President of the Commission of Directors of Libraries of the Council of Rectors of Chilean Universities. He led the commission that generated the Standards for Chilean University Libraries. He has worked as a consultant for international organizations and universities, developing missions in Latin America and Europe. He has been a Visiting Professor in the master's in information sciences programs of Polytechnic University of Valencia (UPV) and Pompeu Fabra University since 2010 and 2008, respectively. He has extensive practical and teaching experience in the various aspects of university management and scientific production.

Héctor G. Ceballos is the Head of the Scientometrics Office of Tecnológico de Monterrey, where he developed its current research information system (CRISTec) and currently advises the development of the Institutional Repository while he conducts analysis on scientific collaboration and research groups' performance. He is also a Collaborator Researcher of the Intelligent Systems group at the School of Engineering and Sciences of Tecnológico de Monterrey, a member of the Masters in Intelligent Systems' faculty staff at Campus Monterrey, and an adherent member of the Mexican Society on Computing (AMEXCOMP). His research lines are semantic web technologies, Bayesian networks and causality applied to research asset management, cultural heritage visualization, and smart cities.

Félix de Moya Anegón is a doctor in philosophy and letters from the UGR, Spain (1992). He is the Founder of Scimago and Principal Investigator of the Associated Unit of the Scimago Group. This unit carries out research and development (R&D) projects, such as the Ibero-American Ranking of Research Institutions (RI3), Scimago Journal & Country Rank, Scimago Institutions Rankings, and the Atlas of Science project for great expansion in Latin America. In recent years, he has worked extensively as a consultant for international organizations such as the World Bank and other institutions. His research activity has led him to sign numerous collaboration agreements with different Spanish entities and other institutions in Chile, Colombia, and Mexico serving for the production of reports on the internationally visible scientific production of different aggregates. In recent years, he has worked extensively as a consultant for international organizations: the World Bank, responsible for the project of analysis of research results in the Balkans and the strengthening of the Technological Science System of Colombia.

Emilio Delgado López-Cózar holds a PhD in library and information science from the Universidad de Granada. He is a Full Professor of Research Methods at the Universidad de Granada (UGR). As a Founder of the EC3 Research Group, he has developed tools for scientific evaluation: Journal Scholar Metrics, Scholar Mirrors, Publishers Scholar Metrics, Book Publishers Library Metrics, IN-RECS, IN-RECJ, IN-RECH (impact factor of Spanish journals in the social sciences, legal sciences, and humanities), H Index Scholar, Proceedings Scholar Metrics, Co-Author Index, I-UGR Ranking of Spanish universities, RESH, EC3 Metaranking of Spanish Universities, and H Index of Spanish Journals according to Google Scholar Metrics. He has authored more than 200 publications, and his lines of research focus on the evaluation of scientific performance (authors, institutions, journals) and Google Scholar as a tool for scientific evaluation.

M'hamed el Aisati is a Vice President of Product Management, Funding, and Content Analytics at Elsevier. He studied computer sciences at the University of Amsterdam before joining Elsevier in 1998. In his early years at Elsevier, el Aisati contributed to the realization of the first Elsevier digital journal platform preceding ScienceDirect and, later in 2004, stood at the cradle of Scopus, which he helped launch. A few years later, el Aisati became the Head of New Technology and in 2010 took on the role of Director of Content and Analytics. He heads a technology and analytics team in assembling relevant data, supporting large research performance evaluation programs, and applying custom analytical tools in large-scale projects that aim to provide insights to governments, research institutions, funding agencies, and university ranking providers. Recently, his responsibilities expanded to include research funding solutions. el Aisati holds two patents and specializes in product development, Big Data, analytics, and service-oriented architecture.

James Fangmeyer Jr. is the Competitive Intelligence for Research Data Scientist at Tecnológico de Monterrey in Monterrey, Mexico. His work is *MoneyBall* for universities, mining hidden value in collaboration networks and other resources to help researchers make a better scientific impact. He has been a Fox Leadership Fellow at the Catholic Volunteer Network and Georgetown University's Center for Applied Research in the Apostolate, where he evaluated international community service programs. Fangmeyer graduated from the University of Pennsylvania's Wharton School with minors in mathematics and statistics. He is the oldest of nine siblings and a lover of the Spanish language.

Nathalíe Galeano is a Director of Competitive Intelligence for Research at Tecnológico de Monterrey, responsible for analyzing research trends and university rankings and generating internal reports for developing research strategies. As a Director of the Research Internships Program at Tecnológico de Monterrey from 2010 to 2013, Galeano integrated undergraduates from different majors into research groups with faculty on campus as well as abroad through collaborative research exchanges. Galeano received her bachelor's degree in production engineering in 1999 from EAFIT University, Colombia; and the title of master in science in manufacturing systems in 2002 from Tecnológico de Monterrey. She has participated as a researcher in projects funded by the European Community within the framework programs FP5 and FP6 in topics related to collaborative networked organizations. Her areas of research interest include collaborative networked organizations, open innovation, entrepreneurship- and technology-based companies, and recently, Scientometrics and scientific and technological foresight.

Shadi Hijazi is a Consultant and a Senior Analyst in the QS Intelligence Unit (QSIU). He has been leading and contributing to consulting engagements in universities around the world, with a particular focus on performance improvement in strategy, internationalization, research, and marketing and branding. His work has included projects with universities in the Middle East, Russia, Malaysia, the Philippines, Korea, and Mexico. He also writes analytical reports for benchmarking university performance and conducts quality assurance audits (Quacquarelli Symonds [QS] Stars). He has been teaching with the London School of Marketing since 2010. His courses there cover diplomas from the Chartered Institute of Marketing, Communication, Advertising, and Marketing Foundation, and the Digital Marketing Institute (DMI). He developed the course materials for courses including digital marketing essentials, digital marketing planning, and implementing digital marketing campaigns. He has also trained professionals in management, marketing, and business fundamentals in various contexts. Hijazi is a Certified Digital Marketing Trainer from the DMI. He holds a PhD in business administration from Kobe University in Japan. His study there addressed the interaction between business performance and knowledge management tools. He investigated the quality of information systems from the point of view of the sales department, and his PhD dissertation focused on the training methods used by Japanese companies to create a knowledgeable sales force.

Frans Kaiser is a Senior Research Associate at the Center for Higher Education Policy Studies (CHEPS), University of Twente, the Netherlands. He is involved in several international research projects funded by the European Commission, studying reforms in higher education and dropout and completion rates in higher education. From 2012 to 2017, he was part of the secretariat that supports the Review Committee—an independent body that monitored the outcomes of the performance contracts agreed with individual higher education institutions in the Netherlands. He is currently working on the implementation of a multidimensional ranking of universities worldwide (U-Multirank), with a special interest in indicators and data at the institutional level.

Bárbara S. Lancho-Barrantes is a Postdoctoral Researcher in Scientometrics at the Tecnológico de Monterrey. She holds a PhD in Bibliometrics with a Bachelor's Degree Extraordinary Award and PhD Extraordinary Award, all obtained from the University of Extremadura. Her scientific production focuses on the field of Bibliometrics, more specifically on scientific citation flows and their relation with the impact indicators. She has published in first quartile journals such as *Scientometrics*, the *Journal of Information Science* (*JIS*), and the *Journal of the Association for Information Science and Technology* (*JASIST*). Moreover, she has presented papers at well-known international congresses of the area: ISSI and COLLNET, III International Seminar on LIS Education and Research (LIS-ER); and on a Spanish level, Federación Española de Sociedades de Archivística, Biblioteconomía, Documentación y Museística (FESABID) and Red de Bibliotecas Universitarias Españolas REBIUN Workshop, and so forth. Her academic and research career has mainly developed within the research group Scimago. Her professional experience includes belonging to top-level Spanish research institutions such as Centres de Recerca de Catalunya (CERCA), Spanish National Research Council (CSIC), and the University of Malaga (UMA).

Nian Cai Liu is a Professor and Dean of the Graduate School of Education of Shanghai Jiao Tong University (SJTU). Liu took his undergraduate study in chemistry at Lanzhou University of China. He obtained his master's and doctoral degrees in polymer science and engineering from Queen's University at Kingston, Canada. He worked as an Associate Professor and a Full Professor at the School of Chemistry and Chemical Engineering of SJTU from 1993 to 1997, respectively. He moved to the field of higher education research in 1999. Liu is now the Director of the Center for World-Class Universities (CWCU) of SJTU. His current research interests include world-class universities and research universities, university evaluation and academic ranking, and research evaluation and science. Liu has published extensively in both Chinese and English. The Academic Ranking of World Universities, an online publication of his group, has attracted attention from all over the world. Liu has been enthusiastic in performing professional services. He is one of the Vice Presidents of IREG-International Observatory on Academic Ranking and Excellence, an international association for the field of university evaluation and ranking.

Alberto Martín-Martín is a University Professor Training Research Fellow (FPU2013/05863 granted by the Spanish Ministry of Education, Culture and Sports) and a PhD candidate in the field of Bibliometrics and scientific communication at the Universidad de Granada. His earlier degrees in library and information science are from the same university, where he graduated with honors. He is currently a member of the EC3 Research Group, where he has collaborated on various research projects, technical reports, and journal articles since 2013.

Wim J. N. Meester received his MSc and PhD in chemistry from the University of Amsterdam. He spent 2 years at Harvard Medical School as a postdoc, before making the change from being an Active Researcher to becoming a Program Manager for the Netherlands Organisation for Scientific Research. In 2006, Meester joined Elsevier as a publisher for the animal science and forensic science journal programs. In 2010, he joined the Scopus team, where he is now responsible for Scopus content strategy and how it meets the needs of customers in the academic, government, and corporate market segments. In addition, he manages the independent, international Scopus Content Selection and Advisory Board and strategic partnerships with relevant research organizations and third-party publishers.

Enrique Orduna-Malea holds a PhD in documentation from the Universitat Politècnica de València, where he currently works as a Postdoctoral Researcher. He belongs to the EC3 Research Group at Universidad de Granada and Trademetrics Research Group at UPV. He specializes in web metric methods applied to the processes of creation, diffusion, and consumption of content and products on the web, both in academic and industrial environments.

Maria-Soledad Ramírez-Montoya is a Researcher and Director of Postgraduate and Continuing Education at the Tecnológico de Monterrey's School of Social Sciences and Education. She is the Coordinator of the Research and Innovation Group on Education, Director of the International Council for Open Education: "Latin America's Open Education Movement," and Director of the UNESCO Chair: "Open Educational Movement for Latin America." She is also a Research Professor invited by the University of Salamanca, where she advises on theses and teaches courses in the Doctoral Program Training on Knowledge Society. Her lines of research include the open educational movement, innovative teaching strategies, technological resources for education, and training of educational researchers.

David Reggio is the Global Head of QSIU Consulting Services. He has extensive experience in scientific and innovation advisory. Among other appointments, he has worked on scientific R&D agendas with the CNPq, Brazil, and the Wellcome Trust, London. Reggio has lectured widely and advised institutions and industry globally on innovation, research, and strategic alliances. He is an elected member of the Royal Institute of Foreign Affairs, Chatham House, London; a Fellow of the Royal Society of Arts & Commerce; and a senior advisor to the Microgravity Space Centre in Rio Grande do Sul, Brazil. With QSIU, Reggio advises on scientific partnership strategies, frontier science and innovation, technological parks, and other priority areas that drive national and international development. Reggio holds a PhD from Goldsmiths College, University of London. His research concerned diagnostic innovation in psychiatry and neuroscience. He subsequently held visiting fellowships at the Oswaldo Cruz School of Public Health, Rio de Janeiro; Seoul National University Hospital; and the University of London.

Cameron A. Ross is a Vice President of Product Management within the Research Products group at Elsevier. Ross holds a BA (first-class honors) degree in politics and Russian studies from Keele University in the United Kingdom and spent part of his curriculum in St. Petersburg in Russia. He was a Territory Sales Manager at Thomson Reuters before joining Elsevier in 2004 as Scopus Sales Manager, where he soon became the Head of Scopus Product Management. In his current role as a Vice President, Ross is responsible for defining the vision and leading the overall Scopus strategy. Based in Amsterdam, the Netherlands, Ross leads a team of more than 20 product managers who work closely with information specialists, data scientists, and engineers to create Scopus, the world's largest abstract and citation database of peer-reviewed literature.

Duncan Ross has been a data miner since the mid-1990s. He currently leads the data science team at Times Higher Education (THE), where he is responsible both for the creation of the World University Rankings and for building out data products to support universities in their development journey. Before this, he worked as a Director of Data Science at Teradata, where he developed analytical solutions across a number of industries, including warranty and root cause analysis in manufacturing and social network analysis in telecommunications. Ross is the Chair of Trustees of DataKind UK, a not-for-profit charity that hopes to improve the entire not-for-profit sector by helping them make the most of the Big Data revolution. They do this by encouraging volunteer data scientists to work for not for profits through short- and longer-term engagements. This work has covered a huge range of organizations, most recently looking at how to use open government data to track the beneficial owners of companies. In his spare time, Ross has been a City

Councilor, a Chair of a national children's charity, a Founder of an award-winning farmers' market, and a member of the UK Government's Open Data User Group.

Joshua D. Schnell is a Director at Clarivate Analytics, with expertise in science planning and assessment, in the evaluation of R&D programs, and in science and technology (S&T) policy; he helps to set the strategic direction of Clarivate solutions. Previously, he oversaw a team of analysts conducting program evaluations of research and training programs and developing tools for data-driven science management. Before joining Clarivate Analytics, Schnell directed the analytics group at a science management start-up in the Washington, DC area, worked in research administration at Northwestern University, and was an S&T Policy Graduate Fellow at the U.S. National Academies of Science. He holds a PhD from Northwestern University.

Michiel Schotten is National Research Assessments Manager at Elsevier, based in Amsterdam, the Netherlands. In this role, he is involved with facilitating the usage of Scopus data both for national research assessments, such as the Research Excellence Framework 2014 in the United Kingdom and Excellence in Research for Australia 2015, and for institutional rankings such as THE and QS. In previous roles, he also contributed to setting up policies and procedures for the Scopus journal evaluation process. In addition to his work at Scopus, Schotten performs research at Leiden University on the echolocation and communication behavior of wild dolphins and for that purpose has constructed a diver-operated recording device to accurately record and localize ultrasonic dolphin sounds (see http://dolcotec.org/projects/4-ch-UDDAS). He holds an MSc in marine biology from the University of Groningen and has also performed research at the University of Hawaii.

Ben Sowter is a leading expert on comparative performance in international higher education. He founded and leads the QSIU, a world's leading think tank on the measurement and management of performance in universities. Since its establishment in 2008, QSIU has extended the rankings to 5 regions and over 40 subjects, as well as innovating in other dimensions. The Intelligence Unit also operates the QS Stars rating system as well as a variety of thought leadership reports on student and applicant behavior, jobs and salary trends, performance of higher education systems, attractiveness of cities as student destinations, and performance of young universities. As a by-product of much of this work, QSIU also provides distinctive business intelligence to over 200 client institutions worldwide and operates two major international conferences: the EduData Summit each June (http://www.edudatasummit.com) and Reimagine Education each December ((http://www.reimagine-education.com). Sowter is frequently invited to international conferences as a speaker and to visit institutions to explain the technicalities behind QSIU work or to draw on the unique data to provide a fresh perspective on trends and developments in international higher education. He has visited hundreds of universities and spoken on our research in over 30 countries in the last decade. Sowter was also appointed to the Board of QS Quacquarelli Symonds Ltd. in 2015.

Susanne Steiginga is a Product Manager for Scopus content at Elsevier, is working on the execution of the Scopus content roadmap and strategy through content expansion programs and new Scopus content releases. She is in close contact with customers, editors, and publishers and is a Scopus liaison for marketing and sales. After completing her bachelor's degree in health sciences, Steiginga continued her academic studies and received two biomedical master degrees in infectious diseases and public health research. After writing her final thesis at the Harvard School of Public Health, she initially started a PhD, which was quickly swapped for a corporate career at Elsevier seven years ago. Steiginga is based in the Amsterdam headquarters, the Netherlands.

Benjamín Vargas Quesada is a University Professor at the Faculty of Communication and Documentation, Department of Information and Communication, UGR, Spain. His lines of research include visualization and evaluation techniques of scientific documentation, visualization and analysis of large scientific domains through network PathFinder (PFNET), social network analysis, classification of science, and stem cell studies. He holds a PhD in documentation and scientific information from the UGR, Spain.

Yan Wu is an Assistant Professor at the CWCU of SJTU. She has been in the ranking team at CWCU since 2005. She obtained her bachelor's (2000) and master's degrees (2003), both in philosophy, from East China Normal Universities. Since 2013, she has been a PhD candidate majoring in public management of education at SJTU. Her primary research interests include the ranking and evaluation of universities. She had been responsible for the Global Research University Profiles project, which started in 2011 and has been developing a database on the facts and figures of around 1,200 research universities in the world. She has spoken on her research at the IREG-7 Conference, and so forth.

Nadine Zeeman is a Research Assistant at the Center for Higher Education, Technical University of Dortmund (TU Dortmund). She works on the project Interdisciplinarity in German Universities. She is also involved in student supervision and teaching. Before joining the Center for Higher Education at TU Dortmund, she worked as a Research Associate at the Center for Higher Education Policy Studies (CHEPS) at the University of Twente, the Netherlands. At CHEPS, she was involved in the projects U-Map, U-Multirank, and EUniVation. Her areas of research interest include policy analysis, research governance and management, transparency tools, and the relationships between universities and society.

Chapter 1

Data Analytics and Scientometrics: The Emergence of Research Analytics

Francisco J. Cantú-Ortiz

Tecnológico de Monterrey, Mexico

Contents

1.1 Introduction

Since its inception in the late 1960s, the Internet has grown exponentially in the number of devices attached to the network all over the world. In October 1969, the first two computers, one in Los Angeles and other in Stanford, California, were interconnected to establish for the first time information exchange between two remote computers (Gromov, 1995). Henceforth, the number of machines connected by networks

1

has increased every year to reach around 3.35 billion Internet users as of November 15, 2015, which represents about 40% of the world population (Internet World Stats, 2017). At the same time, storage capacity has moved from kilobytes (10^3), to megabytes (10^6), to gigabytes (10^9), and nowadays, capacity is measured in terabytes (10^{12}). It is expected that increments within the next few years will require storage measured in petabytes (10^{15}) and in exabytes (10^{18}) as the Internet keeps its pace of growth in both number of users and amount of data generated by these users (Hassan et al., 2010).

This billow of machines and users of the Internet has led to the proliferation of all sorts of data concerning individuals, institutions, companies, governments, universities, and all kinds of known objects and events happening everywhere in daily life. Initiatives such as the *Internet of Things* have arisen in the past few years, making more evident the need for a careful treatment of the tons of data generated from traffic and transactions on the Internet (Vermesan & Friess, 2013). This tendency has led to treating the issue of Big Data, a set of technologies developed to handle huge amounts of data popping up from routine operations of things, persons, and organizations (McAfee & Brynjolfsson, 2012). At the same time, the gems and diamonds residing within data are being discovered through Data Analytics techniques by shedding light over the nooks and crannies of data entanglements and data-link skeins present in the quarry of data (Liebowitz, 2014).

Scientific knowledge is not an exception to the data boom. As the number of academicians and innovators swells, so do the number of publications of all types, yielding outlets of documents and depots of authors and institutions that need to be found in Bibliometric databases. These databases are dug into and treated to hand over metrics of research performance by means of Scientometrics that analyze the toil of individuals, institutions, journals, countries, and even regions of the world. League tables of research performance solely based on research performance have cropped up thanks to various agencies and ranking companies in the last few years.

Research intelligence pursuits compare metrics of achievement to produce rankings of various types, benchmarking dossiers, or scientific trend statements to offer feedback to those actors responsible for assessment of outcomes, appraisal of throughput, or strategy amendment. Thus, *Research Analytics* is understood as the confluence of Data Analytics applied to scientific products found in Bibliometric databases and using Scientometrics and Research Intelligence methodologies in both prospective and prescriptive ways (De Bellis, 2009). In the remainder of this chapter, we elaborate more on each of the elements mingled in the notion of Research Analytics and comment on the main issues around them. Then, the various chapters of the book plow and delve into these elements.

1.2 Data Analytics

As pointed out, Big Data has emerged to deal with large datasets that traditional data processing and relational databases cannot handle appropriately. Multiple commercial platforms and techniques to handle Big Data applications are found now in the market of information technologies. Big Data operations include analysis, capture, data cleaning, search, sharing, storage, transfer, visualization, querying, and information privacy.

To help organizations with Big Data processing, *Data Analytics*, which has become a catch phrase, brought techniques from fields like Artificial Intelligence, Machine Learning, Data Mining, and Statistics, and collected them into a carte of methods and tools to assist users in dealing with the growth of data and information. Data Analytics has called the attention of business, government, and institutions, as it provides useful solutions in digging through Big Data sets to discover patterns and other useful information and knowledge for decision-making. Data Analytics technologies and techniques are widely used in organizations to enable them to make more informed business decisions and by data analysts and researchers to verify or disprove scientific models, theories, and hypotheses.

For most organizations, Data Analytics is a challenge. Data collected across the entire organization come in different formats and are both structured and unstructured representing great difficulties for the whole organization. However, when implemented appropriately, Data Analytics helps institutions to better understand the information contained within data and identify the knowledge that is most important to its strategies (Liebowitz, 2014).

The book series on Data Analytics Applications, of which this book is a part, delves into the ample sorts of domains in which Data Analysis is being used assiduously in a mounting and creative number of ways in fields like sports, health, commerce, education, government, communications, and many others. This book explores the use of Data Analysis in Research, especially in institutions devoted to scientific research and innovation, and in particular, universities and research centers, in an age of competitiveness, globalization, and transformation into knowledge economies.

1.3 Bibliometrics

We have indicated that one area of current data growth is in the sphere of scientific publications. Publications such as journal articles, conference papers, books, patents, reviews, notes, and others are typically stored in Bibliometric databases. A study conducted by Lutz Bornmann and Rüdiger Mutz revealed that the number of papers stored in the Clarivate Analytics's Web of Science (WoS) Bibliometric database grew exponentially between 1980 and 2010, moving from around 700,000 documents in 1980 to nearly 2 million documents in 2010. Additionally, they discovered that the number of references and citations had increased exponentially in the period from 1650 to 2000 (Bornmann & Mutz, 2015). Based on these findings, it has been observed that the amount of scientific knowledge measured in the number of published documents in Bibliometric databases doubles every 9 to

15 years, and this is a trend that will remain for the next few years. This sprouting expansion poses a challenge for most institutions that lack the means or formal mechanisms to adequately document the inception of intellectual contributions taking place within themselves.

Bibliometric databases are of various types. They may be proprietary products, like Clarivate Analytics's WoS (formerly owned by Thomson Reuters) and Elsevier's Scopus, or they can be public products, such as Google Scholar (GS), ResearchGate, and others entering into the Open Access movement. Also, other databases that hold institutional repositories or current-research information systems (CRIS) are reservoirs of publications used by a growing number of organizations. To give a sense of scale, as of September 3, 2014, WoS held around 50,000 scholarly books, 12,000 journals, and 160,000 conference proceedings with 90 million total records, and 1 billion total cited references. Also, 65 million new records are added per year. Chapter 2 reports on WoS. Elsevier, a large research publisher and digital information provider, publishing over 2,500 scientific journals, launched Scopus in 2004. As of 2017, Scopus covers 67 million items drawn from more than 22,500 serial titles, 96,000 conferences, and 136,000 books from over 7,500 different publishers worldwide. Chapter 3 provides an introduction to Scopus. Google inaugurated its GS academic database in 2004. Although GS was initially conceived as a discovery tool for academic information, its designers soon realized the potential GS had as an open-access Bibliometric tool. Chapter 4 describes the general features of the GS search engine in terms of document typologies, disciplines, and coverage, as well as some Bibliometric aids like Google Scholar Metrics and Google Scholar Citations, based on the GS database, and others like H Index Scholar, Publishers Scholar Metrics, Proceedings Scholar Metrics, Journal Scholar Metrics, and Scholar Mirrors developed by third parties. Institutional repositories are technological platforms of open access to knowledge, which are directed to the storage, preservation, and diffusion of the production generated in institutions (MacIntyre & Jones, 2016). Chapter 5 outlines the main platforms used by the 2,824 institutional repositories listed in OpenDOAR as of February 7, 2017.

1.4 Scientometrics

Scientometrics applies Data Analytics methods and tools to Bibliometric databases to calculate scientific indicators in science, technology, and innovation. What Scientometric methods and tools can help us understand Bibliometric databases and the trends they contain? Several lines of research have developed to answer this question. These include how to understand scientific citations; how to measure scientific impact, including that of researchers, journals, and institutions; how to compare scientific disciplines from a Scientometrics standpoint; and what kinds of indicators should be used in assessment, policy, and management of research. Other research issues include normalization to make unbiased comparisons, the mapping of scientific fields, and among many other countries. Scientometrics also permits studies about research collaboration, hot research topics, research trends, patenting, funding, and other related topics (Cantú & Ceballos, 2012). Thus, Scientometrics measures progress in science, technology, and innovation over time (Price, 1978).

The field of Scientometrics is based on the work of its cofounders Derek J. de Solla Price (1922–1983) and Eugene Garfield (1925–2017). Price, a British physicist, formulated his theory of the exponential growth of science in the 1940s by studying the Philosophical Transactions of the Royal Society. He noticed that the growth of these transactions between 1665 and 1850 followed a log linear trajectory, one of the characteristic signs of an exponential process (Price, 1963).

Garfield, an American linguist and businessman, created the *Science Citation Index*, which encompasses WoS, the *Journal Citation Report* (JCR), and the Impact Factor (IF), and he founded the Institute for Scientific Information (ISI) in 1955 (Garfield, 1955). JCR evaluates the world's leading journals with quantifiable, statistical information based on citation data. The impact factor of a journal is a measure reflecting the yearly average number of citations to recent articles published in that journal. It is frequently used as a proxy for the relative importance of a journal within its field. Journals with higher impact factors are often deemed to be more important than those with lower ones. Impact factors are calculated yearly starting from 1975 for those journals that are listed in the *Journal Citation Reports*. In 1992, ISI was sold to the company Thomson Reuters, which sold it in 2016 to Clarivate Analytics, and their metrics are widely used for Scientometric analysis. Chapter 2 describes the Scientometric approach used in WoS.

Other metrics were developed after the indices proposed by Garfield, including the H Index of a scientist, where H is the number of publications the scientist has with H or more citations over his/her career, and normalized factors that control for differences in scientific behavior across disciplines, time periods, publication types, or geographic regions. Chapter 3 describes the Scientometric approach utilized within Scopus. Chapter 4 does the same for GS. Finally, the journal *Scientometrics* was established in 1978 to promulgate research findings and cutting-edge academic publications in the field.

1.5 World-Class Universities

The concept of world-class universities was brought forward in the early 2000s by scholars like Altbach (2004, 2005), Alden & Lin (2004), and Levin et al. (2006). Jamil Salmi makes the case that world-class universities are characterized by three distinguishing attributes: highly sought graduates, leading-edge research, and technology transfer (Salmi, 2009). Salmi adds that these attributes are complemented by three sets of factors at play in top universities: a high concentration of talent, including faculty and students; abundant resources to offer a rich learning environment and conduct advanced research; and favorable governance features that encourage strategic vision, innovation, and flexibility and enable institutions to make decisions and manage resources without being encumbered by bureaucracy (Salmi, 2009). Although this sounds fine, in practice, world-class universities are generally recognized by the research they generate and the influence this research infuses into technology development and knowledge-based economies. To understand the circumstance of world-class universities and their profiling, Liu and colleagues inaugurated the Center for World-Class Universities at Jiao Tong University, Shanghai, China, and launched in 2005 the International Conference on World-Class Universities series, which takes place every 2 years (Sadlak & Liu, 2007).

With this context, the longing to compete for a place among world-class universities became evident at the dawn of this millennium among higher-education institutions. This aim has materialized through the publication of big-league tables displaying rankings of world-class universities. Higher-education institutions compete with each other and even sometimes collaborate to attract international students, faculty members, and external funding, and boost reputations and other aspects of world-class universities. The use of World University Rankings (WUR) by talented students who look for a university to pursue undergraduate or graduate studies, academicians who search for a place to start an academic career, companies that establish partnerships with centers of knowledge, and governments that allot funding to institutions or allocate student scholarships has become evident in the last two decades (Hazelkorn, 2015). The main players in WUR, which characterize one aspect of world-class universities, are presented in the next section.

1.6 World University Rankings

World University Rankings (WUR) are big league tables of universities listed from first to last, ordered by a score that is calculated using in most cases a polynomial function whose arguments are university attributes chosen by a ranking designer to reflect aspects of interest of the ranked institutions, and using weights arbitrarily set by the same designer. Averaging, smoothing, normalization, discrimination, and other statistical techniques are often used to account for biases or disparities when comparing disciplines and geographical regions, or dealing with missing data.

The age of publishing international big league tables displaying rankings of top universities was initiated at Jiao Tong University in China by N.C. Liu and Y. Cheng, who lifted the banner that inaugurated the rankings era by publishing the Academic Ranking of World Universities (ARWU) for the first time in 2003 (Liu & Cheng, 2005). Remarkably, the ARWU has been the most stable ranking in the last decade because its methodology has not changed. The methodology favors elite universities with a particular set of funding, facilities, and faculty because its indicators include the number of Nobel Prize-winning faculty and alumni, the number of articles published in *Nature*, and the number of articles published in *Science*, among others. The ARWU methodology is explained in Chapter 6.

This was followed by the publication of a ranking by Times Higher Education (THE) and Quacquarelli Symonds (QS), which joined efforts to issue the THE-QS World University Ranking from 2004 to 2009. In 2010, THE and QS parted ways, with QS keeping and refining the methodology they owned since 2004 with an accent in employment indicators. THE developed its own methodology in 2010, emphasizing the industry–university relationship as a criterion, among others.

QS is a higher education specialist company that provides rankings, guides, and services. It identifies four main pillars that contribute to a world-class university: research, teaching, employability, and internationalization. To measure these pillars, QS collects data from university data submissions, the Scopus database, and reputation surveys of academics and employers. The QS WUR methodology is explained in Chapter 7.

THE is a ranking agency that became independent in 2005 of *The Times*, a newspaper based in London, England. Its ranking focuses on research universities that offer a broad curriculum of undergraduate education. Data are submitted by universities, requested in reputation surveys, and extracted from the Scopus database. Metrics from these data sources are organized into the five main categories of the ranking: teaching, research, citations, industry income, and international outlook. THE World University Rankings methodology is explained in Chapter 8.

Scimago Labs is a technology-based company offering innovative solutions to improve the scientific visibility and online reputation of institutions. It is based in Madrid, Spain, and is responsible for the design and publication of the Scimago Institutions Rankings (SIR), Scimago Journal & Country Rank (SJR), and Shape of Science, an information visualization project whose aim is to reveal the structure of science based on journal citations. The SIR and SJR methodologies are explained in Chapter 9.

A key player in the use of nontraditional metrics is the Webometrics Ranking of World Universities, also known as the Web Ranking of World Universities. It is an international ranking that uses alternative metrics (Altmetrics)—a variant of traditional metrics that use citations to publications—to produce a composite indicator that takes into account both the volume of web content (number of web pages and files) and the visibility and impact of these web publications according to the number of external inlinks (site citations) they receive. Webometrics is published by Cybermetrics Lab, a research group of the Spanish National Research Council Centro Superior de Investigación Ceintifica (CSIC) based in Madrid, Spain. The aim of the ranking is to improve the presence of the academic and research institutions on the web and to promote the open-access publication of scientific results. The ranking started in 2004 and is updated every January and July. Today, it provides web indicators for more than 24,000 universities worldwide. The Webometrics methodology is explained in Chapter 10.

U-Multirank is a transparency tool that allows users to create their own rankings. U-Multirank was launched in 2014 based on five main design principles:

user-driven, multidimensional, comparability, multilevel nature of higher education, and methodological soundness. Data come from institutional and student questionnaires and the Web of Science database. The U-Multirank methodology is explained in Chapter 11.

Although *U.S. News & World Report* has published university rankings since the early 1980s, these only included U.S. universities, excluding higher-education institutions from other countries. *U.S. News & World Report* started publishing its Best Global Universities (BGU) ranking in 2014 to join the globalization game of higher education. This ranking does not solicit data submission from universities. It draws data from reputation surveys and the Web of Science database. The indicators used by *U.S. News & World Report* are designed with a focus on research. The U.S. News Best Global Universities methodology is explained in Chapter 12.

We include in Section II of this book the world university rankings that are the leaders in the field. Although these are not exempt from discussions and controversies, they have attained credibility for their methodology, data curation, comparability, and stability over years (Bowman & Bastedo, 2011). However, there are other less-known rankings, among which are the the the following: (a) Leiden Ranking, produced by the Centre for Science and Technology Studies at Leiden University, of European and worldwide universities ranking in the top 500 according to the number and impact of Web of Science (WOS)–indexed publications per year; (b) the Round University Ranking, a world university ranking published by an independent agency located in Moscow, Russia, assessing the effectiveness of 750 leading universities in the world based on 20 indicators distributed among 4 key dimension areas: teaching, research, international diversity, and financial sustainability; and (c) the University Ranking by Academic Performance (URAP), developed in the Informatics Institute of Middle East Technical University, which has been publishing annual national and global college and university rankings for the top 2,000 institutions. The Scientometrics measurement of URAP is based on WoS.

Other rankings include the uniRank™ (formerly 4 International Colleges & Universities or 4icu.org), which is an international higher education search engine and directory reviewing accredited universities and colleges in the world that includes 12,358 colleges and universities, ranked by web popularity, in 200 countries; and the UI GreenMetric World University Ranking, an initiative of Universitas Indonesia, launched in 2010 to provide the result of an online survey regarding the current condition and policies related to Green Campus and Sustainability in the Universities all over the world. There exist many other regional and national rankings published periodically, which lie outside the scope of this study (Basu, 2013).

To assess, evaluate, and certify the various ranking methodologies, the International Ranking Expert Group (IREG) was founded in 2004 by the UNESCO European Center for Higher Education (UNESCO-CEPES) in Bucharest, Romania, and the Institute for Higher Education Policy in Washington, DC (Audit, 2011; Cheng & Liu, 2008). It became the IREG Observatory on Academic Ranking and Excellence (IREG Observatory), a nonprofit organization, in 2009.

IREG runs several initiatives that include the Berlin Principles of rankings and league tables of higher education institutions and programs, the IREG Ranking Audit, the IREG Inventory of National Rankings, the IREG List of International Academic Awards, and the IREG Guidelines for Stakeholders of Academic Rankings. These initiatives are described in the IREG website. The Berlin Principles was promulgated in IREG's second meeting held in Berlin, Germany, from May 18 to May 20, 2006. The meeting was convened to consider a set of principles of quality and good practice in university rankings. It is expected that

> It is expected that this initiative will set a framework for the elaboration and dissemination of rankings—whether they are national, regional, or global in scope—that ultimately will lead to a system of continuous improvement and refinement of the methodologies used to conduct these rankings. Given the heterogeneity of methodologies of rankings, these principles for good ranking practice will be useful for the improvement and evaluation of rankings. http://ireg-observatory.org/en/index.php/berlin-principles

The ethical issue in the elaboration of world-university rankings, and related themes, is addressed by Sedigh (2016) and further developed in Chapter 13.

1.7 Research Analytics

Research Analytics studies the behavior of scientific and innovation activities in institutions and applies Scientometric analysis to assess progress, do institutional benchmarking, conduct intelligence studies, and develop internal key process indicators (KPIs) to identify main strengths and weaknesses and tune institutional strategies. Research Analytics is supported by a framework of methods and tools that utilize Data Science and Bibliometric databases to do Scientometric analysis of scientific research, innovation, and technology transfer in universities and research institutions. Research Analytics is an emerging multidisciplinary area that combines methods and tools from Artificial Intelligence, Machine Learning, Statistics, Computer Science, and the Social Sciences to measure the impact of science and technology. Scientific knowledge has been growing exponentially during the last few decades, doubling the number of publications every 9 to 15 years. Citations follow a similar pattern. Data Analytics methods and tools are being applied to Bibliometric databases like Web of Science, Scopus, Google Scholar, and institutional repositories to discover new knowledge and find relationships among data.

In the near future, Research Analytics, understood as the synergy between Data Analytics, Bibliometrics, Scientometrics, and research intelligence in both prospective and prescriptive ways, will become even tighter to deal with the rapid growth of scientific papers worldwide (Downing & Ganotice, 2017). The chapters of this book present the main Bibliometric databases and Scientometric methods in use to measure research performance as well as the leading world university ranking methodologies and how they use Bibliometrics and Scientometrics to produce world-class university rankings.

References

Alden, J., & Lin, G. (2004). Benchmarking the characteristics of a world-class university: Developing an international strategy at university level. *Leadership Foundation for Higher Education*, London.

Altbach, P. G. (2004). The costs and benefits of world-class universities. *Academe, 90*(1), 20.

Altbach, Philip G. (2005). A world-class country without world-class higher education: India's 21st century dilemma. *International Higher Education, 40*, 18–20.

Audit, IREG-Ranking (2011). Purpose, criteria and procedure. IREG Observatory on Academic Ranking and Excellence.

Basu, A. (2013). World university rankings. *SRELS Journal of Information Management, 50*(5), 679–690.

Bornmann, L., & Mutz, R. (2015). Growth rates of modern science: A bibliometric analysis based on the number of publications and cited references. *Journal of the Association for Information Science and Technology*.

Bowman, N. A., & Bastedo, M. N. (2011). Anchoring effects in world university rankings: Exploring biases in reputation scores. *Higher Education, 61*(4), 431–444.

Cantú, F. J., & Ceballos, H. G. (2012). A framework for fostering multidisciplinary research collaboration and scientific networking within university environs. *Knowledge Management Handbook: Collaboration and Social Networking*, pp. 207–217.

Cheng, Y., & Liu, N. C. (2008). Examining major rankings according to the Berlin principles. *Higher Education in Europe, 33*(2–3), 201–208.

De Bellis, N. (2009). *Bibliometrics and citation analysis: From the science citation index to cybermetrics*, p. 417. Scarecrow Press.

Downing, K., & Ganotice, F. (2017). *World university rankings and the future of higher education*. Hershey, PA: IGI Global.

Garfield, E. (1955). Citation indexes for science. *Science*. American Association for the Advancement of Science, *122*(3159): 108–111.

Gromov, G. (1995). Roads and crossroads of the Internet history. Netvalley.com

Hassan, S., Humaira, & Asghar, M. (2010). Limitation of silicon based computation and future prospects. *Second International Conference on Communication Software and Networks, IEEE Computer Society*.

Hazelkorn, E. (2015). *Rankings and the reshaping of higher education: The battle for world-class excellence*. Springer.

Internet World Stats (2017). http://www.internetworldstats.com/stats.htm.

Levin, M. H., Jeong, D. W. & Ou, D. (2006). What is a World Class University? Conference of the Comparative and International Education Society, Honolulu, HI.

Liebowitz, J. (ed.) (2014). *Business analytics: An introduction*. Boca Raton, FL: CRC Press.

Liu, N. C., & Cheng, Y. (2005). The academic ranking of world universities: Methodologies and problems. *Higher Education in Europe, 30*(2, July): 127–36.

McAfee, A., & Brynjolfsson, E. (October 2012). Big data: The management revolution. *Harvard Business Review.*

MacIntyre, R., & Jones, H. (2016). IRUS-UK: Improving understanding of the value and impact of institutional repositories. *The Serials Librarian, 70*(1–4), 100–10.

Price, D. J. de Solla (1963). *Little science, big science.* New York: Columbia University Press.

Price, D. J. de Solla. (1978). Editorial statement. *Scientometrics, 1*(1).

Sadlak, J., & Liu, N. C. (2007). *The world-class university and ranking: Aiming beyond status.* Bucharest: UNESCO-CEPES.

Salmi, Jamil (2009). *The challenge of establishing world-class universities.* Washington, DC: World Bank.

Sedigh A. K. (2016). Ethics: An indispensable dimension in the university rankings. *Science and Engineering Ethics, 23*(1), 65–80.

Vermesan, O., & Friess, P. (2013). *Internet of things: Converging technologies for smart environments and integrated ecosystems.* Aalborg, Denmark: River Publishers.

SCIENTOMETRIC DATABASES

Chapter 2

Web of Science: The First Citation Index for Data Analytics and Scientometrics

Joshua D. Schnell

Scientific and Academic Research, Clarivate Analytics, Rockville, MD

Contents

2.1 Introduction

In 1955, Dr. Eugene Garfield published an article in *Science* magazine titled "Citation Indexes for Science: A New Dimension in Documentation through Association of Ideas" that laid out the concept of an index of scientific publications that used the authors' own references as a way to identify ideas contained within a scholarly article (Garfield, 1955). This insight led to the *Science Citation Index*, published first in print in 1964 and then online as the Web of Science in 1997. The *Science Citation Index* and the subsequent decades of innovation brought by Garfield, his colleagues at the Institute of Scientific Information (ISI), and their collaborators around the world served as the foundation for the burgeoning Scientometric community in Europe and the United States. The various successors to the ISI, now Clarivate Analytics, have developed and released new datasets that extend and enhance Garfield's pioneering vision of a citation index that reflects the scientific enterprise.

Now the Web of Science includes the *Science Citation Index Expanded* (SCIE), the *Social Science Citation Index* (SSCI), the *Arts & Humanities Citation Index* (AHCI) as well as the recently released *Emerging Sources Citation Index* (ESCI). Today's Web of Science has evolved to reflect the reality that today's scientific enterprise relies heavily on bridging disciplinary divides and recognizing and incorporating research produced in emerging scientific communities. The globalization of the Web of Science has been accompanied by a globalization in the practice of Scientometrics, particularly in emerging regions where rapidly growing government investment in science and technology is leading to an increased use of quantitative methods for identifying and rewarding research excellence.

This chapter is written to provide the reader with an introductory understanding of the history, composition, and uses of the Web of Science citation databases for Bibliometric and Scientometric analysis. I start with the genesis of the *Science Citation Index* and the evolution of citation indexes from information retrieval to Scientometrics. Next, I provide a description of the principles and practicalities of source selection and coverage of the Web of Science, followed by recent enhancements that facilitate Bibliometric and Scientometric analysis, including the *ESCI* and improvements in author disambiguation. This is followed by a brief treatment of the meaning and theories of citation behavior and the use of citation-based measures as indicators of research influence and impact. Then, I describe the use of citation databases for monitoring and mapping research trends and close the chapter with a review of the current literature that reflects the use of Web of Science as a research resource.

2.2 Genesis of the Web of Science: Garfield and the Foundation of Bibliometrics

The origins of the *Science Citation Index* (SCI), the earliest incarnation of what we now call the Web of Science, came from attempts to overcome the challenges associated with information retrieval of the scholarly literature. As described by Nicola de Bellis in the thorough treatment of the subject in his 2009 book, *Bibliometrics and Citation Analysis*, librarians and information scientists in the 1950s and 1960s were looking for ways to index the growing scholarly literature using the automatic processing technologies of the day (primarily computers using information inputted through punch card) (De Bellis, 2009). However, the natural language-based techniques of the time proved to be technically impractical to achieve the accuracy necessary for a robust indexing approach. Dr. Eugene Garfield, who had moved from chemistry into information science, applied the insight that an author records conceptual links to other scholarly and nonscholarly documents within the bibliography of a scholarly journal article. He recognized that these links were the scientific publishing equivalent of the citation links that had been used for decades for information retrieval in the legal field (*Shepherd's Citations*).

In his 1955 paper in *Science*, Garfield described this insight in detail, providing an example from a single research article that helped to describe his new indexing process. Garfield called this approach an "association-of-ideas index," and it, in his view, would bring together work that normally wouldn't be connected or even appear in the field-specific indexes of the day. He also introduced the benefit of a collection of citations of an article of interest in providing an indication of the influence, or impact, of a paper on the scholarly discourse and publishing of the time. Garfield's articulation of the "impact factor" of the paper was the start of a half century of innovation in tracking the evolution of ideas through citation indexes.

It was approximately 10 years later, after several pilot studies and work on an initial Genetics-focused citation index, that Garfield released the *SCI* in 1964. He and his colleagues added the *Social Science Citation Index* in 1972 and the *Arts & Humanities Citation Index* in 1978. These citation indexes provided the data that the Scientometrics community needed to begin to test at a larger scale many of the Scientometric theories that had been developed in the preceding decades. The *SCI* data were used to carry out statistical analysis of the literature and provided a foundation for the development and maturation of the Scientometric community.

With the rapid adoption of the World Wide Web in the academic community in the 1990s, Garfield's citation indexes were brought online in 1997 as the Web of Science. The availability of the Web of Science as an Internet-based platform, the adoption of electronic publishing platforms, and the expansion of quantitative management into academia all led to the adoption of Web of Science as the *de facto* tool for research evaluation. This new application as an assessment tool was added to its primary and secondary purposes, the former as an information retrieval tool and the latter as the only data source for large-scale science of science studies.

It was the mid-2000s before any competing citation indexes appeared, and the emergence of these competitors (Scopus, Google Scholar) signaled a new era in the adoption of citation-based Scientometrics for evaluation of research by administrators and researchers themselves. For the last dozen years since alternative citation indexes have appeared, the Scientometric community has worked to better understand how these different tools reflect scholarly communication and the process of conducting science. Repeatedly, researchers find that the fundamental principles of quality of the Web of Science, described further in the next chapter, continue to hold true in this era of rapid expansion of publication and globalization of scientific research.

2.3 Selection of Content and Editorial Standards

The Web of Science Core is composed of the three primary Citation Indexes (the *SCIE*, the *SSCI*, and the *AHCI*) and the *ESCI*, which was created in 2015 to apply the same editorial standards of the three flagship indexes to a broader set of rapidly proliferating journal titles in emerging disciplines and emerging world regions. With over 1 billion cited reference connections indexed from high-quality, peer-reviewed journals, books, and proceedings, it is the only scholarly literature database that provides consistent, controlled, and curated indexing for all authors and addresses, cited references, and funding acknowledgements. Today, *SCIE* includes over 8,800 journal titles with 48 million records, *SSCI* over 3,200 journal titles with over 8.5 million records, *AHCI* over 2,500 journal titles with over 4.6 million records, and *ESCI* with over 5,000 emerging peer-reviewed journals. In total, 65 million records and the resulting 1 billion cited reference connections are annotated with high quality metadata ensuring that Scientometricians can create optimal analysis groups for their project, from individuals to universities, from regions to subject areas, and through time.

From its beginnings the Web of Science has been developed using a consistently applied editorial process that emphasizes the selection of the world's best journals. The value of selecting the core of the best journals was based on the notion of Bradford's Law, initially published by S. C. Bradford in 1934, which suggested that for a given scientific discipline, the core literature was composed of the papers in a small set of journals, and that in relatively few one could find the most important papers for that discipline (Bradford, 1934).

Garfield extended this notion further by articulating the principle known as "Garfield's Law of Concentration," which states that the core literature for all scholarly disciplines is concentrated within a small set of core journals, the set of which changes as the nature of research in specific disciplines changes too (Garfield, 1971).

These two principles, Bradford's Law and Garfield's Law, have served as the basis for the journal selection process employed for the Web of Science. As the literature evolves, the Web of Science changes as well. An in-house team of subject experts identifies and evaluates promising new journals, and decides to discontinue coverage of journals that have become less useful or whose standards have dropped. With the introduction of the *ESCI*, journals that have not yet fully met the high threshold for selection of the three flagship indices can be indexed and covered in the Web of Science, providing exposure to the global academic community until that time as they are incorporated into the main indexes.

Before covering in detail the editorial process and principles of the Web of Science, it is important to emphasize that it is a *complete cover-to-cover citation index*. That means that during the indexing process, the editorial teams use a combination of computerized and human indexing tools to capture the complete journal issue contents from cover to cover. Other scholarly databases select only certain document types to capture.

In addition, all cited references of a paper are indexed. Those cited references that have already been indexed in the Web of Science (so-called source items) are resolved to the original record, and for those that are not resolved, a new record is established. As there can be repeated citing instances of a cited reference that is not a source item, these cited references records are also resolved, enabling users of the Web of Science to measure the citation impact of documents that are not part of the core journal sets. This cover-to-cover indexing and the resolution of cited references ensure that the Web of Science provides the highest-quality data, which is especially important for Data Analytics and Scientometrics.

2.3.1 Editorial Process

Journal selection for the Web of Science is an ongoing process and journals can be added or removed at any point. In a typical year, the editorial team will evaluate roughly 3,500 titles for possible inclusion in the three main collections, with approximately 10% of these candidates ultimately selected. The editors are also continuously monitoring journals that have already been selected to maintain quality and editorial consistency across the different Web of Science editions. The editorial team considers many factors when selecting journals, including whether the journal meets basic publishing standards, whether the journal has an international focus, the quality of the editorial content, and analysis of the citation patterns of the journal title and its articles.

2.3.1.1 Basic Publishing Standards

The most important publishing standards to be considered are ethical publishing practices and quality peer review. A robust peer-review process not only bolsters the quality and veracity of the research study being published but also ensures the completeness and quality of bibliographic data included. A publisher that exhibits predatory or unethical editorial practices that may lead to fraudulent citation behavior, such as citation cartels, will not be added to the Web of Science, and the title will be deselected when appropriate.

Another important standard to be considered is timeliness. Journals under consideration must demonstrate that they are meeting their stated publication frequency, which indicates that they have a sufficient backlog of manuscripts for ongoing viability. Journals that publish regular issues must show reliable timeliness for a minimum of three issues, while publishers that post articles individually online are reviewed after a minimum of 9 months of content production.

Additional publishing criteria include the publishing format, which can be either print or electronic format (PDF or XML); English language publishing in full text, or at least abstracts; and standard editorial conventions that assist with information retrieval, such as descriptive article titles and abstracts, complete bibliographic information for all cited references, and full address information for each author of an article.

2.3.1.2 International Focus

It is important for a journal that is under consideration to appeal to a large international audience; therefore, the editorial team looks for evidence of international diversity among the journal's editorial advisory board members, editors, and contributing authors. In the case of excellent regional journals, editorial staff will deemphasize the requirement for international focus and instead look for complementarity of journal content to the titles already within the relevant Web of Science collection.

2.3.1.3 Editorial Content

The editorial team members are constantly assessing their area of science or the arts to identify emerging disciplines and topics that may become the subject area of new journal titles. When a new title is under consideration, there is an assessment of how well this title's content fits with existing content within the subject area. It is important for a journal under consideration to enrich the database and not to significantly duplicate existing content.

2.3.1.4 Citation Analysis

Because the Web of Science is made up of citation indexes, the editors also leverage the citation data within the various databases to determine the influence and impact of journals that are being evaluated for coverage. There are two primary indicators that are used: Total Citation (TC) counts and Journal Impact Factor (JIF). These two indicators provide the editors with a better understanding of how a journal contributes to the overall historical citation activity within the scholarly community in a size-dependent way (TC) and the recent average impact of the journal in a size-independent way (JIF).

Citation data are considered within the overall editorial context of the journal, including looking at the citation data of editors and authors, as well as the general citation dynamics within the journal's subject area. The phenomenon of self-citation is normal; however, the *Journal Citation Report* (JCR) editors monitor the level of self-citation that occurs within a journal's overall citation activity. Excessive rates of self-citation are examined further, with possible suppression or deselection of the journal if citation patterns are found to be anomalous.

2.3.2 Emerging Sources Citation Index

In recognition of the growing importance of local and regional scholarly literature, the *ESCI* was launched in 2015 to deepen the coverage of Web of Science and make the scholarly output of emerging fields and regions more visible. By design, *ESCI* is a multidisciplinary citation index covering all areas of the sciences, social sciences, and arts and humanities. Journals that are under evaluation for inclusion in *ESCI* are expected to conform to many of the same editorial standards as the rest of the Web of Science, including evidence of strong peer-review and ethical publishing practices, publication of literature of high interest to the scholarly community, implementation of minimum technical requirements necessary for indexing, English language bibliographic information, and recommendations for inclusion from the scholarly community of users of the Web of Science.

Journals that are under consideration for inclusion in one of the three main indexes may first be added to the *ESCI* based on the discretion of the editorial team. In addition, journal titles that are deselected from the three main indexes may also be covered in *ESCI* if the editorial development staff deem this appropriate. However, selection for *ESCI* does not ensure that a journal title will be added at some point to the main citation indexes. Additionally, journals in the *ESCI* are not duplicated in one or more of the flagship citation indexes.

An initiative to retrospectively index up to 10 years of the back issues of journal titles accepted into the *ESCI* was started in 2017. This added content will provide users with the historical depth necessary to leverage this content for information retrieval and Scientometric purposes.

2.4 Web of Science Features Important for Scientometrics and Data Analysis

As a foundational source of data for the Scientometric community, the Web of Science has features that enable the creation of Scientometric indicators and robust data analysis. In this section, recent features that are of most interest to Data Analytics will be covered, including the use of subject categories, enhancements for organizational and author disambiguation, and the recently developed Item Level Usage Metrics (ILUM).

2.4.1 Subject Categories in the Web of Science

Although it is difficult to precisely specify the subject of scientific and scholarly output, subject of field categorization is an essential part of any citation index that supports information retrieval and Data Analytics. Journal titles within the Web of Science are assigned into one or more subject categories, using a process that started with the *Science Citation Index*, and has been managed over the years by the editorial staff based on their expert knowledge of their research domains. The editorial team maintains a set of subject category descriptions, called "Scope Notes," which reflect the content of the journals that have been included in this subject category over the years. It is not intended to be an ideal description of all possible aspects of the research field. To assign a new journal to a subject category, an editor with knowledge of the field will examine (a) the editorial content of the journal under evaluation for topical relevance and (b) the citation patterns of the journal in terms of what journals and fields cite it and what journals and field its cites. Based on these two relevance indicators, a journal will be assigned to one or more subject categories.

The scope notes for categories are reviewed continually and updated as the content of the journals included in the category shifts, and likewise the assignment of journals into these categories is assessed regularly. As scientific and scholarly fields expand and/or undergo specialization, our editors make determinations to create new subject categories and to reassign journals as appropriate.

It is important for the data analyst to recognize that journals can be assigned to multiple subject categories. As described previously, citation patterns vary by subject area, so the subject categories within the Web of Science are used to generate field- or category-normalized article metrics. If a journal title is assigned to multiple categories, it is necessary for the analyst to review the field-normalized indicators for all fields to determine the overall relative influence of those articles across related but distinct subject categories.

2.4.2 Unification of Organization Variants

When using the Web of Science as an analytic data source, it is often important to aggregate publications (the typical unit of measurement) accurately in multiple ways, for example, by subject area, or by author organization or geographic region. Subject-area aggregation is achieved by assignment of journals to subject categories as described previously. Organizational or regional grouping is made difficult by the variety of ways authors report their institutional affiliations. Over the last few decades, author affiliations within the Web of Science have been processed and unified in order to permit accurate summaries by organization or region. To do this, a team of data analysts works continually with publishers, research institutions, and authors to map all present forms of an affiliation and address to a parent organizational record. This mapping, called Organization-Enhanced, enables the user to search at the level of the unified parent organization and at the level of the specific variant. Organizations are also categorized into organization type, such as Academic, Government, and Corporate, thus facilitating organizational-based comparisons and collaboration analyses.

2.4.3 Author Identification Enhancements

One of the most difficult challenges facing Scientometricians when working with the scholarly literature is *author name disambiguation*. Historically, many scholarly publishers recorded only a last name and the initial or initials for each of the authors on a publication. Identifying unique researchers among those with the same last name and initials can be difficult without additional information about the authors. Starting in 2008, the Web of Science began to index the full name forms and linked each name to the researcher's institutional affiliation and address.

However, with the volume of data available in the Web of Science, large-scale disambiguation of author names requires more than just an author name and affiliation. Users can take advantage of three features to resolve ambiguity in author names: the Web of Science integration with two different identification systems, ResearcherID (RID) and Open Researcher & Collaborator ID (ORCID), and the Distinct Author Identification System (DAIS) in the Web of Science.

ResearcherID is a free tool that was created in 2008 to allow registered users to associate Web of Science papers with a research profile. ORCID is a similar identification system that was established in 2012 by an independent foundation using the ResearcherID technology. ORCID also allows scientists to create a scholarly profile populated with research publications and other scholarly activities. In 2016, ORCIDs identifiers were integrated into the Web of Science so that publications that have been claimed by ORCID users are now discoverable using these unique identifiers. Both of these IDs provide human-verified data for address and author name disambiguation.

The Web of Science also takes advantage of a system of proprietary algorithms, called DAIS, to automatically resolve possible publications to individual authors. This algorithmic approach uses up to 25 different weighted elements of an article (e.g., author e-mail or citing author lists) to evaluate the likelihood that an author of a pair of articles is the same person. The papers are then assigned to the author's unique ID. This entirely machine-driven approach also takes into account the presence of a ResearcherID and any publications claimed by the authors associated with this ID.

2.5 Citation-Based Measures as Indicators of Research Impact

The creation of the *Science Citation Index* came at an important time for historians and philosophers of science who had spent the preceding decades conceptualizing and refining theories of the dissemination of scholarly knowledge and debating how well those theories reflected the social norms that underpinned the scientific enterprise. The newly available citation data Garfield produced allowed a cadre of academics to begin to test these theories. Garfield and his collaborators subscribed to the "normative" theory of citation, promoted by Dr. Robert K. Merton, a sociologist of science at Columbia University. This school of thought followed the conceptualization of citations as expressions by an author of intellectual debt to those scientists who came before. Merton viewed scientists as subject to a moral imperative to cite the work of others when participating in the formalized process of publishing a research report. Thus, when an author references prior work in the bibliography of her research paper, she extends credit to the authors of these prior works as contributing to her own knowledge base and scientific understanding. For those research papers that are cited repeatedly by subsequent authors, there is an implicit recognition that the value of the work is greater than those papers that are cited less frequently, or not at all. In other words, the work of a scholar that receives more citations, and thus more credit, compared to her scholars working at the same time in publishing in the same journals and in the same field is inherently more influential and therefore more valuable. This view of the value of a research paper being reflected in the number of citations to the paper in subsequent years became the standard construct of the *impact* of research that was adopted by Bibliometricians and Scientometricians.

In the decades following the release of the *Science Citation Index*, a number of citation-based measures were developed and released, the most visible of which is the *Journal Impact Factor* (JIF). The JIF was created by Garfield and Irving Sher in order to provide a citation-based view of the impact of a journal that would help in selecting journals for inclusion in the *Science Citation Index*. Over the years, it has become an important indicator in the scientific publishing community, and it is one of approximately a dozen metrics that are calculated for journals annually in the *Journal Citation Reports* (JCR). The JIF counts all citations in a given year to all of the papers published in the journal in the previous 2 years and then divides that total by the number of research articles in that journal in the previous 2 years. It can be considered a measure of the central tendency of a journal's papers to be cited in subsequent years, and it has become a standard tool for understanding a journal's influence and impact on the scientific community. Within the JCR, journals are ranked within their subject categories, enabling the scientific community to assess where journals fall in terms of overall impact when selecting a venue for publishing research.

The JIF has not been without controversy, as it has been incorrectly applied as a stand-in for the impact of a specific article, or articles, and at times as a proxy measure of the quality of a researcher's work. In response to this misuse of JIF, a number of new indicators have been proposed and developed, with recent advances leading to article-level measures that provide a more accurate view of the influence of a particular research paper, or for the collection of papers authored by a scientist. Today, using Web of Science citation data, Scientometricians are able to compare the citations to a given article published in a given year to the average citations received by all other articles published in the same year in the same journal (journal-normalized citation impact) or to the average citations received by all other articles published in the same year in the journals that are in the same subject category (field-normalized citation impact). These normalized measures are the best choices for analysts looking to assess the impact of a set of articles by a research group, department, or institution.

2.6 Mapping Science Using the Web of Science

Since Derek de Solla Price first proposed turning the tools of science on science itself (de Solla Price, 1963), measuring and mapping the scientific enterprise using the scholarly literature in the Web of Science has been a desire of policy makers, researchers and Scientometricians. Beyond simply counting the papers published in specific journals or subject categories of Web of Science, the citation relationships that have been comprehensively indexed for decades allows for a clustering of papers to represent the real structure and dynamics of specialty areas and, when aggregated, domains of investigation. Often analysts will follow the paths of direct citation through subsequent generations of papers to map the evolution of our understanding of, say, a given disease or physical phenomenom. However, Scientometricians can also use the methods of *bibliographic coupling* and *co-citation* to reveal more about how scholarly research actual forms, ebbs and flows, grows or dies.

In the 1960s, Myer M. Kessler of Massachusetts Institute of Technology recognized that two or more publications that reference many of the same works in their bibliographies share a cognitive coupling and therefore could be considered to be concerned with similar topics (Kessler, 1963). When using this method, termed *bibliographic coupling*, the strength of the similarity depends on the number of shared cited references. Bibliographic coupling, however, is static since the references in a paper's bibliography are determined at the time of publication. Thus, the couplings that are produced are not updated in any way as our understanding of the science related to these works inevitably advances.

To address this limitation, Henry Small, a historian of science working with Garfield at ISI, realized that papers frequently cited together by newly published literature was another method to define closely related papers produced by members of the same "invisible college" (Small, 1973). The method, termed *co-citation*, provides a forward-looking view of emerging subspecialties as new publications appear that introduce new co-citation pairs and reinforce existing co-citation pairs. As with bibliographic coupling, the frequency of the co-citation pairs indicates the strength of the topical similarity between the publications.

Co-citation was chosen to serve as the basis of the science mapping efforts through the years, led by Small. Research Fronts, specialty areas determined by co-citation analysis, are produced every 2 months in Essential Science Indicators, part of the Web of Science (Small, 2003). These Research Fronts are clusters of papers published in the last 5 years that are cited in the top 1.0% when compared to papers in similar journals and subject categories published in the same year. Core papers in the Research Front are the frequently co-cited papers forming the foundation of the specialty; however, it is the citing papers, on the leading edge of research, that define the relationships and strength of connections that create the Research Front. The mean publication year of the core papers in the cluster can be used to identify those Research Fronts that have emerged more recently, whereas the total number of citations to core papers can identify those Research Fronts with the largest influence.

Research Front analysis has uses in research administration, funding, and policy making. An analyst interested in understanding where a university stands among its peers and how it is performing in terms of active and emerging research can use that organization's presence in Research Fronts clusters as one indicator. Such an approach has been used since 2002 by Scientometricians at Japan's National Institute for Science and Technology Policy (NISTEP). Using Research Fronts data, the NISTEP team has produced biennial maps for policy makers to understand how Japan's research community is participating in and contributing to emerging research clusters (Saka & Igami, 2014).

2.7 Analysis of Recent Publications Using Web of Science as an Analytic Resource or as the Focus of a Research Study

For close to half a century, the Web of Science has been a critical resource for the discovery of important literature, for evaluation of the impact of journals, and the productivity of research organizations and scientists, often within a subject area. In more recent years, the Web of Science has also become an analytic resource for researchers interested in using the citation histories contained within it as a proxy for large-scale analysis of the knowledge flows in the scientific enterprise especially in the context of networks or graph theory.

In the last 20 years, computing power has dramatically increased while data storage prices have rapidly declined. This opportunity has led many universities and research organizations recently to establish data science and Big Data Analytics research programs. Big Data Analytics has proven to be a compelling new research area for social scientists, behavioral economists, and researchers interested in the "Science of Science." While the "Science of Science" is not a new research discipline, it has been reinvigorated in the past decade as national governments around the world have emphasized the need for evidence-based policy making and have established systematic support for the development of new ways to understand and manage national scientific systems, particularly by funding the development of new empirical methods that leverage Big Data resources to study science at scale. The Web of Science, with the most complete historical coverage and citation linkages, has been increasingly used as a dataset for researchers trying to understand the scientific enterprise as it is reflected in high-quality, peer-reviewed scholarly literature.

To get an initial look at the use of Web of Science as a research resource, on December 20, 2016, we performed an analysis of all publications indexed in the Web of Science that include the exact phrase "web of science" in the Title, Abstract, Author, Keywords, or Keyword Plus fields (Topic search). This yielded a total of 11,869 records, with 5,948 (50.1%) of those categorized as Reviews and 5,455 (46.0%) categorized as Articles. Remarkably, 44.3% (5,258) of the records were published in the last 2 years (2015 or 2016), and the last 5 years accounted for almost 80% of the records. This signals an explosion in the use of the Web of Science as a research resource and not just as a tool to discover important literature within the research areas indexed.

Looking at the most frequently occurring Web of Science categories, we observed that the primary research areas represented by these publications are biomedicine and health, information sciences, and computer sciences. Concentration of the use of the Web of Science as a research resource in these subject areas reflects the acceptance of meta-analysis and systematic reviews in the health and biomedical sciences as an established research paradigm, and it is consistent with the increased development and application of data science methods to the citation index data within the computer science and information science fields.

In order to understand more carefully the research that relies heavily on the use of the Web of Science citation indices, we restricted the above publication set to those that include the exact phrase "web of science" within the Title field only, resulting in a set of 282 records. Of these, 194 (69.2%) publications are categorized as Articles, and 117 (41.7%) are in the WoS category of "Information Science & Library Science." We selected publications of Article document type only to conduct a customized categorization based on the use of the Web of Science in the article.

Five of the articles were deemed not relevant due to the topic, and were excluded. Of the remaining 189, the following was the topical breakdown:

∎ 3 articles deal with the "Science of Science" or the structural basis of scientific inquiry as represented by scholarly publishing trends.

- 37 are Scientometric studies, including studies of metadata captured in citation indices, indicators developed using these databases, or comparisons of indicators across different data sources.
- 31 are related to information retrieval using citation databases, one of which specifically relates to the use of Web of Science data to inform university library collection management.
- 17 involve research assessment exercises primarily focused on academic institutions, departments, and researchers, and many compare the Web of Science to other citation databases.
- 101 are specific Bibliometric analysis papers in particular research areas, regions, or countries, which use the Web of Science as a data source to understand the landscape of research in a particular subject area or location and to determine the impact of that research.

Of the 189 articles reviewed, 57 (30.2%) include a comparison to other citation databases, primarily Scopus (45 articles) or Google Scholar (25 articles). Of the 67 articles that concern a particular geographic location, the most common country analyzed is Spain (20 articles), followed by Brazil (7 articles). There are 107 (56.6%) articles focused on a particular subject area, and Psychology (18 articles) and Neuroscience (18 articles) are the most frequent subject areas analyzed.

Papers published in the mid-2000s are focused on information retrieval, followed by papers published in the late-2000s being primarily focused on Scientometrics and the comparison of the Web of Science to alternative citation databases. From the late 2000s to the present, articles documenting Bibliometric analyses of particular subject areas are most prevalent, with Scientometric studies being fairly constant at a lower level.

Within the small subset of articles studied here, the most frequent use of the Web of Science is as a research resource in Bibliometric analyses of particular subject areas and/or regions. A larger analysis of the full set of articles identified here is necessary to determine if articles using Web of Science to study the scientific enterprise are increasing as a result of increased attention from governments and funders around the world.

2.8 Concluding Remarks

As additional data sources for Big Data Analytics and Scientometrics are developed and released, there is an ever-growing risk to novice and experienced practitioners alike: namely, that the lure of more and bigger data will overshadow the importance of meaningful, high-quality data. To mitigate that risk, the Web of Science will continue to focus on select, curated content in line with the founding principles of Dr. Eugene Garfield when he launched the *Science Citation Index* in 1964 and the *Journal Citation Reports* in 1975. Clarivate Analytics, the current custodian of the *Science Citation Index*/WoS heritage, continues carefully to expand and enhance data, indicators, and tools for the next generation of scholars and scientists, Scientometricians, and science evaluators and policy makers.

References

Bradford, Samuel Clement. "Sources of Information on Specific Subjects," *Engineering*, 137(3550): 85–86, 1934.

De Bellis, Nicola. *Bibliometrics and citation analysis*, Maryland: Scarecrow Press, 2009.

Garfield, Eugene. "Citation Indexes for Science: A New Dimension in Documentation through Association of Ideas," *Science*, 122(3159): 108–11, 15 July 1955.

Garfield, Eugene. "The Mystery of the Transposed Journal Lists—Wherein Bradford's Law of Scattering Is Generalized According to Garfield's Law of Concentration," Current Contents #17, in *Essays of An Information Scientist 1962–1973*, pp. 222–223. Philadelphia: ISI Press, August 4, 1971.

Kessler, Myer M. "Bibliographic Coupling between Scientific Papers," *American Documentation*, 14(1): 10–25, 1963.

De Solla Price, Derek J. *Little Science, Big Science*, New York: Columbia University Press, 1963.

Saka, Ayaka & Masatsura Igami, "*Science Map 2010 & 2012—Study on Hot Research Area (2005–2010 and 2007–2012) by Bibliometric Method*," National Institute of Science and Technology Policy (NISTEP REPORT No. 159), July 2014.

Small, Henry. "Co-citation in Scientific Literature—A New Measure of the Relationship between Two Documents," *Journal of the American Society for Information Science*, 24(4): 265–269, 1973.

Small, Henry. "Paradigms, Citations, and Maps of Science: A Personal History," *Journal of the American Society for Information Science and Technology*, 54(5): 394–399, 2003.

Chapter 3

A Brief History of Scopus: The World's Largest Abstract and Citation Database of Scientific Literature

Michiel Schotten[1,2], M'hamed el Aisati[1],
Wim J. N. Meester[1], Susanne Steiginga[1],
and Cameron A. Ross[1]

[1]*Research Products, Elsevier B.V., Radarweg, Amsterdam, The Netherlands*
[2]*Institute of Biology Leiden, Leiden University, Sylviusweg, Leiden, The Netherlands*

Contents

Imagination is more important than knowledge.

Albert Einstein
1879–1955

In these most interesting of times, in which basic phenomena such as the size of crowds or the link between human greenhouse gas emissions and global warming are often disputed and regarded as a matter of personal opinion, there is a pressing need for a source of verifiable and authoritative information. While modern science, sparked by the scientific revolution and standing on the shoulders of giants such as Copernicus, Galileo, and Newton, is by no means perfect and has many shortcomings, it is arguably also one of the most successful human endeavors in history: since its inception in the sixteenth century, modern science has not only led to incredible progress in standards of living and technical breakthroughs for humanity, but also provides us with a clear method to obtain reliable, verified information that we can trust. And now, such a trustworthy source of information is more important than ever before: where matter was still the primary currency until fairly recently, this was replaced by energy at the start of the Industrial Revolution—as formalized by Einstein's famous equation $E = mc^2$, which states that matter is just another form of energy—and with the Internet revolution at the start of the twenty-first century, we have now clearly entered the Information Age, with information in turn replacing energy as the most important currency (Jahn, 1996; Verlinde, 2011, 2016; Wheeler, 1992). An abundance of information, however, can easily lead to information overload. While any Internet search engine or social media forum will yield a plethora of data points on any topic imaginable, it is precisely the *trustworthiness* of this information that often comes into question and that we need to base our decision-making on. And while information obtained through the scientific method, which has been peer reviewed and vetted, is arguably the most trustworthy that we have available, we do need a tool to navigate the vast ocean of research information out there. Enter Scopus, the largest curated abstract and citation database of peer-reviewed research literature in the world.

Scopus is still relatively young and a newcomer in this space: it was founded only in 2004 by a small team at Elsevier in Amsterdam, the Netherlands. Elsevier itself is a large research publisher and information analytics company, publishing over 2,500 scientific journals such as *The Lancet* and *Cell*; their name has been derived from the Elzevir publishing house in Leiden in the seventeenth century, which in 1638 published Galileo's final work, *Discorsi e Dimostrazioni Matematiche Intorno a Due Nuove Scienze*, thus saving it from the Inquisition of the Catholic Church (de Heuvel, 2017). When Elsevier launched Scopus, the odds were not very much in its favor: at the time, most universities around the world already subscribed to another major abstract and citation database (covered elsewhere in this book), which was already well established and known for its infamous *Impact Factor* (IF) rating of scientific journals. However, in its young history, Scopus has done incredibly well so far, growing from a mere 27 million indexed items in 2004 to over 67 million items at the time of writing in 2017, drawn from more than 22,700 serial titles; 98,000 conferences; and 144,000 books from over 5,000 different publishers worldwide, and last year, in 2016, it introduced its own *CiteScore* metric of journal performance. Today, Scopus is being used as the primary research citation data source by researchers and health professionals from top universities and research institutes around the globe, by leading university ranking organizations such as Times Higher Education (THE) and Quacquarelli Symonds (QS), by funding organizations such as the European Research Council (ERC) and the National Science Foundation (NSF), and by evaluation bodies conducting national research assessments such as the Research Excellence Framework (REF) in the United Kingdom in 2014 and the Excellence in Research for Australia (ERA) exercises in 2010, 2012, and 2015. This chapter aims to give an overview of Scopus' remarkable success story since its inception, including its user interface, functionality, advanced analysis tools, and Bibliometric indicators; its coverage of different subject areas, publication years, sources, and document types; its stringent content selection policies and processes; the many author and affiliation profiles that can be found in Scopus; how Scopus data feed into other Elsevier research products such as SciVal, ScienceDirect, Pure, and Mendeley; and how customers use Scopus Custom Data (SCD) and Scopus application programming interfaces (APIs) to incorporate into their own tools.

3.1 Launch and Early History of Scopus

The idea for Scopus arose in response to needs expressed by users, in discussions with librarians and researchers from more than 20 of the world's leading research organizations. These included the need to be able to search research papers across many different scientific disciplines (as the majority of abstract databases at the time were based on content within a single discipline), the ability to link citing and cited documents both forward and backward in time, and last but not least, the ability to access the full text of those papers, at least from those journals that an institution subscribed to. Thus, building of the all-science database Scopus started in 2002 and in March 2004, a first beta version was released for extensive user testing by this group of librarians and researchers. Based on their feedback, including feedback on Scopus' intuitive user interface (in an approach known as *user-centered design* [UCD]), the product design was fine-tuned and Scopus was commercially launched in late 2004 on https://www.scopus.com (Boyle & Sherman, 2005); in addition, a U.S. patent was granted to the underlying technology (Zijlstra et al., 2006).

The name *Scopus* was conceived during a team outing in the woods: it was named after the African bird *hamerkop* (*Scopus umbretta*), a bird species so unique that it earned its own genus *Scopus* and that is known (besides for making huge nests of 1.5 meters across that can hold a grown-up man) for its outstanding navigation skills—similar to how Scopus helps researchers navigate the world of scientific literature. In addition, *Scopus* is also the name of an ornithology journal (which can actually be found in the Scopus database), as well as that of a mountain outside of Jerusalem. The word *Scopus* itself originally derives from Latin, meaning "goal" or "target," and even found its place in the philosophy of Spinoza, who used the word *scopus* to indicate a goal that leads to happiness (de Dijn, 1996). The Latin word *scopus* in turn comes from the classical Greek word σκοπός (from the verb σκέπτομαι, "to see") that means "watcher" or "one that watches," as in the word *telescope* (literally "far watcher"). Thus, Scopus provides its users with a unique *view* on the world's research literature, helping them to easily *navigate* it and quickly reach their desired *goal* or destination.

3.2 Scopus' Main Use Cases and Functionality

In its most basic form, Scopus may be considered as a scientific search engine. Type any keyword, phrase, article title, journal name, or author name, and Scopus will almost instantaneously return up to millions of relevant documents. These can easily be sorted and filtered further by publication year, subject area, affiliation or country of the authors, document type, and so forth, or any combination thereof; ten types of refinement filters on the left side of the screen make this job a no-brainer. Search results are presented in a very structured and intuitive overview, showing all the basic and most relevant information of each result: article title, authors, year, and journal name, including links to the full text of the article as well as the number of citations that it received. Clicking on that citation number will immediately show all the articles (that is, only those that are also indexed in Scopus) that have cited the article in question in their References section. And this uncovers one of the greatest strengths of Scopus: the ability to quickly see how often an article has been cited and, more importantly, by whom. This is extremely useful not only to the authors of the article themselves, who want to know who have cited their work and what new research is coming out in their own niche discipline, but also to university provosts, ranking agencies, research funders, and governments of this world; increasingly, citation metrics are being considered as one of the proxies for quality of the published research.

When aggregated among an author; department; research institution; or even a country or group of countries, region, or sector; citation data offer a unique (albeit one-sided) view into that entity's research performance—at least in terms of the "splash" that it is making among its peers, that is, how the published research is being taken up, used, and referenced in the work of others. As with any metric, however, there are of course caveats and limitations regarding their interpretation and to what extent they may and should be applied—mainly depending on the specific context in which they are used (Colledge & James, 2015).

Scopus offers citation linking both forward and backward in time: any publication indexed in Scopus can be linked both to the publications that came before it, that is, that it cites in its References, and to the publications that came after it and that in turn cite the publication in question. Thus, every record in Scopus provides a snapshot in time of whatever specific research topic that the publication is about: it is very easy to see which previous research that publication's research was based on and which later research has built upon it further, thus offering a unique view on how the research on that topic has evolved over time, branched off into other topics, and so forth. As such, it places every published research finding inside a very large intricate network, where each node may ultimately be connected, through forward and backward citation links, to almost every other publication indexed in Scopus. This also makes searching for literature on any topic a breeze: after typing some initial keywords and zooming in on publications of interest (with the most highly cited ones as prime candidates), a user can then easily click either on the documents that were cited by those publications or on the documents that cite them in turn, thus gaining a very quick understanding and overview of a topic with a minimum of effort and time.

What is more, not only is it possible to do so for the elementary units that Scopus is made up of, that is, metadata records of research publications, but one can also do such citation analyses for larger aggregates of those records: author profiles, affiliation profiles, and scientific journals and books (each of which will be treated in separate sections of this chapter). Thus, a user can easily keep track of the entire body of work of a specific researcher (at least for those publications indexed in Scopus), see who has cited that researcher and how often, how many self-citations that researcher has made, and so forth. Also, the amount of research published by university X within time frame Y, as well as the citation performance of that published research within subject area Z, can quite easily be tracked and plotted—such data from Scopus are also used as input into Elsevier's SciVal tool. This type of information is invaluable to decision makers at universities, who want to know the strengths and weaknesses of their university, in what areas to invest, which top researchers to attract, and so forth. The same goes for research funding organizations, which are keen to track the research output of their investments and how that research is picked up by others and contributes to its field. Also, ranking organizations such as THE and QS use aggregated citation data from universities as one of the pillars on which they base their university rankings each year, and it is very useful to governmental research assessment agencies as well, which need to make decisions on how to distribute the available research funding among a country's institutions.

3.3 Basic Scopus Record

The Scopus search result page allows for easy sorting, either by the number of received citations (with the most highly cited records shown on top) or date (with either the most recent or the oldest records on top), among others. From this search result page, a user can select any record of interest and click either on the article or book chapter title (which will open its basic Scopus record); on any of its authors

(which will open the Scopus author profile of that author); on the journal, conference proceeding, or book name (which will bring up the *Source details* page for that
publication outlet); on the *Full text* or *View at Publisher* links (which will open the
access page to the full article on the publisher's site); or on the *Cited by* link that
was discussed in the previous section. Most often, a user will simply click to open
the basic Scopus record of the article, book, or book chapter (Figure 3.1). This will
show a host of additional information about the publication: besides essential metadata such as its title, authors, publication year, volume, issue, page or article number, and name of the publication outlet, also many other details are shown such as
its *Digital Object Identifier* (or DOI, which is a unique identification code for each
scientific article), the author affiliations (with links leading to the Scopus affiliation
profiles), the document type (e.g., a research article, review, book chapter, or conference paper), source type (e.g., a journal or book), the *International Standard Serial
Number* (ISSN) of the journal or *International Standard Book Number* (ISBN) of
the book, the publisher name, author keywords, possible grant information, and
language of the publication. For those users doing a literature search, the abstract
will be of prime interest, allowing them to get a quick synopsis of what the research
is about and its main findings—and very importantly, and what makes Scopus so
unique as has been discussed: the full list of references of the publication.

Each of the documents listed in the References section and that is also indexed
in Scopus is clickable and links directly to the Scopus record of that document.

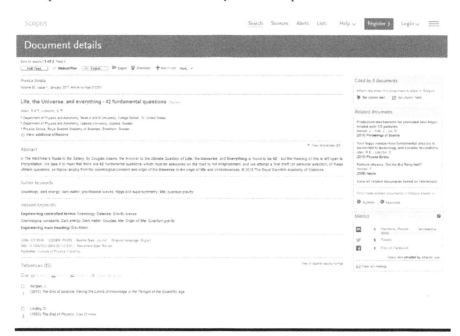

**Figure 3.1   Scopus record of the 2017 review article "Life, the Universe, and
Everything—42 Fundamental Questions" (Allen & Lidström, 2017).**

But the user does not necessarily need to go there, as the *Cited by* link to all the documents citing that specific reference is already present in the References section itself, as are the links to its full text on the publisher's site. As mentioned, whether a reference is clickable depends on whether it is indexed in Scopus, and the way through which this citation link is established is by a robust algorithm that matches the metadata available in the reference to the metadata of the relevant Scopus record. This algorithm is one of the rock solid foundations that Scopus is built on and from which it derives its strength and its countless applications in Bibliometric analyses. Another interesting element for such analyses is the relatively recent addition of a *Metrics* section to the Scopus record, which includes traditional Bibliometric indicators such as the citation count, citation benchmarking, and the publication's Field-Weighted Citation Impact (FWCI). The FWCI is a measure of how well cited an article is compared to similar articles of the same age, document type, and within the same subject area; by definition, the global average FWCI of such similar articles is 1.00, so anything higher means that an article is well cited.

But besides these traditional indicators, the Metrics section also shows newer, so-called alternative metrics, such as the number of times that the document has been added to someone's library on Mendeley (Elsevier's free reference manager) or how often it has been mentioned in mass media, in blog posts, or on social media such as Twitter, Facebook, Google+, and Reddit. If more details are desired, the user can open a link to the *Metric Details* page of the publication, from where, for instance, the specific blog posts, Tweets, or mainstream media articles about the publication can be accessed. Needless to say, this is invaluable information especially to the authors of the publication themselves and showcases how Bibliometrics has expanded far beyond traditional citation analysis in recent years. Another very valuable section of the Scopus record is the *Related documents* section of the record (just above the Metrics section), which shows all the documents in Scopus that have either one or several references in common with the publication, one or several of its authors, or one or several keywords (the user can select which type of related documents is shown). Besides a literature search based on citation links (i.e., retrieving documents that are either cited or were cited by the document in question), this is another very useful way to collect articles of interest on a specific topic. And last but not least, assuming that, in most cases, a user will want to obtain the full text of a publication of interest, or to save it otherwise, the basic Scopus record contains many links to do so: for instance, if the user's institution happens to be subscribed to the publication outlet of the record in question, a *Full text* button is shown, offering immediate access via the publisher's site, or the *Download* link will download the PDF directly. If the user's institution is not subscribed to it, the *View at Publisher* link is an alternative way to link to the publisher's website from where the full text may be purchased or downloaded, or users can choose to save the record's metadata to their personal Mendeley library, add them to a specific bibliography, print or e-mail them, or save them to a self-created list in Scopus for later reference.

A final note on the basic Scopus record concerns its unique identification code: while the DOI uniquely identifies a scientific publication in general (at least from those serial titles and books that are registered with CrossRef, the organization that issues the DOIs), the Scopus record itself also has a unique identifier, called the electronic identifier (EID). This Scopus EID can be found in the URL of the record and starts with *2-s2.0-*; it is used to identify the record in the back-end processes of Scopus. In principle, this EID is a static property of a Scopus record, in the sense that it does not change over time, so that a Scopus record should always be retrievable using its EID—although there is only one scenario where a Scopus record itself may be removed. This happens when two (or in rare cases, several) Scopus records of the same publication are merged into a single Scopus record, for which there may be several reasons. The most common reason is when the record of a so-called Article in Press (AIP), which is a preprint of an article that has been accepted for publication but does not have volume, issue, and page numbers assigned to it yet, is merged with the record of the officially published version of the article, once it comes out. As Scopus indexes AIPs for over 5,100 serial titles from all the major publishers, this is a very common procedure; a matching algorithm running in the back-end of Scopus ensures that such mergences happen automatically and (for the most part) flawlessly. Another reason for merging Scopus records, however, is when it is discovered that Scopus accidentally indexes duplicate records of the same document, for instance, for content indexed via MEDLINE (which is a bibliographic database of life sciences and biomedical journals compiled by the National Library of Medicine in the United States); all MEDLINE content is indexed in Scopus via a direct feed, but for journals that are also sourced directly from the publisher, such MEDLINE content is automatically merged with the Scopus record of the same document, which may in very rare cases lead to duplicate records if the mergence did not happen correctly. For a database with the size of Scopus, consisting of over 67 million records and growing, such mistakes can and do happen on rare occasions—but relatively speaking, this is only a tiny proportion, and effective correction procedures are in place to ensure merging of such duplicates (as well as corrections of any other metadata errors) in the shortest time frame possible, whenever they are noticed.

3.4 Scopus Author Profiles

Besides the basic Scopus record, the two other most elementary units in the Scopus database on which Bibliometric analyses are based are the Scopus author profile and Scopus affiliation profile; the latter two types of units are aggregations of the first type. Scopus author profiles can be accessed either by clicking on an author's name in a Scopus record or by performing an author search, using the last and first name (or initials) and affiliation of the author, or the author's Open Researcher and Contributor ID (ORCID, a unique researcher ID), as search entries. When searching

Figure 3.2   Scopus author profile of famous physicist Stephen Hawking at the University of Cambridge.

on the name *Stephen Hawking*, for example, the Scopus author profile in Figure 3.2 is obtained. The area highlighted in gray shows the most important information, that is, Professor Hawking's full name (as well as his name variants on the right), his current affiliation at the University of Cambridge, and his Scopus author ID, which is a unique identifier for this particular author profile. Immediately underneath, we can see that at the time of writing this, Professor Hawking has 148 publications indexed in Scopus, which together were cited an impressive 29,478 times, yielding him a so-called *h*-index of 68 (which is a researcher performance metric, defined as the number of publications that received at least as many citations; see Hirsch, 2005)—and that he published these papers together with a total of 42 (no coincidence, see Figure 3.1) coauthors. His publications can directly be viewed and accessed underneath, while useful additional information is found on the right: for instance, the *Show Related Affiliations* link (in the *Author History* box) shows all the additional affiliations where Professor Hawking has published in the past, while the chart provides a good overview of his publications and received citations in recent years. Someone interested in following his latest work can click the *Follow this author* button, which sets an e-mail alert to get notified whenever a new publication of his is indexed in Scopus—while Professor Hawking himself could opt to improve his ORCID profile (which happens to be https://www.orcid.org/0000-0002-9079-593X) with the structured information from his Scopus profile, by clicking the *Add to ORCID* link.

ORCID is also the name of the nonprofit organization that registers ORCIDs; its aim is to provide one unique code to each academic author for disambiguation purposes, similar to how the DOI uniquely identifies scientific articles. The

ORCID initiative started in 2012, and at the time of writing, in 2017, over 3 million ORCID profiles have already been registered. Researchers who have created an ORCID profile and wish to populate it with their publications' metadata can easily do so by looking up their profile on Scopus and from there export a complete list of their publications (or at least those that are indexed in Scopus) to their ORCID profile, using the *Add to ORCID* link mentioned earlier. This triggers the so-called *Scopus2ORCID feedback wizard*, which can also be accessed freely (outside of Scopus, for those without a Scopus subscription) at http://www.orcid.scopusfeedback.com. While this process will simultaneously improve their Scopus profile as well, a second, more direct way through which authors can manually curate and update their own Scopus profile is through the *Author feedback wizard*—either by clicking the *Request author detail corrections* link from within the Scopus profile itself or outside of Scopus at https://www.scopus.com/authorfeedback. This easy-to-use tool enables authors to select their preferred name that is shown in the profile—as there may be quite a few variations of their name present in Scopus, for example, written with one or several of their initials or their full first name, a different last name before and after marriage, and so forth; second, to select from the publications that are displayed in their profile, one by one, which of those are their own (and should therefore remain) and which are not and should therefore be removed from their profile; and last but not least, to search for any of their publications in Scopus that may be missing from their main profile but should be added to it.

A significant number of author feedback correction requests are processed automatically, and these automatic corrections to the profiles are usually displayed in Scopus already within two to three days of the request. This is only half the story, though, behind the general high quality of Scopus author profiles: before any manual correction can take place, Scopus employs one of the most sophisticated algorithms in the industry to create such profiles in the first place, fully automated and (in most cases) with a high level of accuracy. This algorithm (called the *Scopus Author Identifier*) clusters documents in Scopus and assigns them to specific author profiles, based on matching of the author name, their e-mail address, affiliation, subject area of the articles, citation pattern, coauthors, and range of publication years, among other parameters. Or, if no match to an existing profile can be found (e.g., when a researcher has just published his or her first paper), it will create a brand new author profile with that researcher's name and only one publication in it. In general, the process of grouping documents and matching them to a specific profile (or creating a new one) goes through a progressive series of steps, or matching filters, whereby more scrutiny is applied (i.e., more matching steps are required) to very common names—think of *Joe Smith*. The idea behind this is to avoid that documents from two or more authors with the same name are accidentally merged into a single profile. This approach reflects the greater value that is placed upon creating profiles with a high *precision* over those having a high *recall*—where *precision* is defined as the relative number of documents in an author's profile that were indeed written by that author (and therefore correctly assigned to the profile) and *recall* refers to the relative number of an author's documents that can be found in their largest profile, if they happen to have more than one. Indeed, our careful approach to avoid mixing up *multiple* researchers named Joe Smith in one profile has the potential flip side that a *single* researcher Joe Smith may end up with several smaller profiles in Scopus, each with only a portion of his publications in it. As long as his largest profile still contains the majority of his publications, though, the recall can still be said to be high.

At the time of writing in 2017, the average precision of author profiles in Scopus is very high at 98% (meaning that, on average, 98% of documents in a profile indeed belong to that author), while the average recall is 93.5% (so that on average, the largest profile of an author contains about 93.5% of that author's total publications, with the other 6.5% spread out over one or several smaller profiles). While these numbers are not perfect, they are quite impressive considering the huge size of the Scopus database, with over 67 million records and over 12 million author profiles, and also considering that they have been algorithmically rather than manually created. One geographical area, though, where still significant improvements can be made, is Southeast Asia: especially some Chinese names, such as Wang or Li, are so common that it could easily happen that several people with both the same given name and family name work not only in the same field, but also at the same university and even within the same department. Thus, even if the most stringent matching rules are applied by the algorithm (as is now standard procedure for Asian names), several researchers with such similar names could still end up within the same Scopus profile. Indeed, such so-called over-merged Scopus profiles are a known issue for Southeast Asian researchers and they drag down the average profile precision statistic for Scopus as a whole—which also implies that the precision is actually 100% for a very large number of Scopus author profiles, namely of those researchers who are not Asian. It should also be noted that Scopus strictly keeps to its policy to index every article exactly as it was originally published, including the exact author names and affiliation names as they appear in each publication. As such, the quality of author and affiliation profiles in Scopus depends to a large degree on the quality of the author and affiliation names in the original publications that they are based on—and there may be many variations in how the name of a particular author or affiliation is written across many different publications.

But the best part of this ongoing narrative is the realization that, as The Beatles already sang, "It's getting better all the time!" As Scopus is getting traction as the de facto standard Bibliometric database to use, an increasing number of researchers realize the importance of having a correct and complete author profile in Scopus (e.g., for grant and tenure evaluations) and will utilize the very easy-to-use tools available to manually check and correct their own profiles—currently, Scopus receives and processes over 250,000 such profile enhancement requests per year. As corrections by the authors themselves can be expected to yield 100% accurate profiles, it means that as time goes by, the precision and recall of Scopus profiles will eventually approach 100%—never quite reach it, of course, as errors in any database of this size remain almost unavoidable, but getting very, very close. And, unlike the other large curated abstract and citation database and Scopus' nearest competitor (which has only a relatively small number of manually curated researcher profiles), Scopus is currently the only database of its sort that applies the "golden combination" of both *algorithmic* profiling of its entire corpus of records en masse and *manual* curation of those automatically created profiles. And as continuous enhancements and improvements are being made to the Scopus Author Identifier algorithm itself as well, this results in ever better profiles with ever higher accuracy and completeness.

3.5 Scopus Affiliation Profiles

The third basic entity in the Scopus database, besides the Scopus record of a publication and the author profile, is the Scopus affiliation profile (which is sometimes also called the institutional profile). In this context, *affiliation* refers to any organization that authors list in their publications as being affiliated to; most often these are universities, but they may also be research organizations, hospitals, corporations, governmental or nonprofit organizations, and so forth. The challenge with profiling affiliations, even much more so than with profiling author names, is that there may be many different name variations of any particular organization, for instance, written in the local language of a country versus its name in English, using abbreviations or acronyms versus the full name or name changes throughout the history of an institute (e.g., before and after a merger with other institutes). What is more, even if only a single name were to be used across all publications from an institute, there may be spelling mistakes in the affiliation string in any one of those publications, or the institute could have multiple addresses and branches in different cities, any one of which could be used in separate publications—and even the country name in the affiliation address could change over time, in case of geopolitical events such as the disintegration of the Soviet Union. Complicating matters even further is that oftentimes, organizations performing research may have subdivisions with a hierarchical structure (such as the parent–child relationship that may exist between a university and a university hospital), where the child organization could also be conceived of as an independent affiliation on its own—or alternatively, both the parent and child organization may be mentioned within the same affiliation string of a publication. As is becoming obvious, automatic profiling of publications based on the affiliation strings is no easy task—but again, Scopus is the only abstract and citation database that does so for its entire corpus of records, using a state-of-the-art algorithm called the *Scopus Affiliation Identifier*.

At the time of writing, over 70,000 different affiliations are searchable in Scopus by typing any name (or part of the name) of an affiliation or of the city or country where the desired affiliation is located. If we want to retrieve the profile of Princeton University in the United States, for instance, we can simply type the word *Princeton* in the Affiliation search field, which currently brings up 73 organizations that either have the word *Princeton* in their name or, if not, are located in Princeton, New Jersey. The most prolific of these is indeed Princeton University, and its affiliation profile shows the number of published documents, authors, and patents from this institute in Scopus (Figure 3.3). We can also see its unique Scopus affiliation ID, as well as the institutes with which it has collaborated (in other words, whose researchers have been coauthors on publications that were also authored by researchers at Princeton) and in which sources (i.e., journals, conference proceedings, or books) Princeton researchers have published most often, with links to the actual indexed publications for each category. On the right, the button *Follow this affiliation* will set an e-mail alert for when new publications from Princeton become indexed, and underneath it, a chart shows which subject areas (associated with the publications) are the most common; it is obvious that Princeton University performs very strongly in the natural sciences, such as physics, mathematics, and engineering.

In a similar fashion as for the Scopus author profiles, the average precision and recall statistics for Scopus affiliation profiles (at least at the time of writing in 2017) can be calculated to be 99% and 95%, respectively. And while providing ad hoc feedback on the correctness of these profiles was already possible, there has not been a structural and automated tool available to do so, that is, at least not one on par with the Scopus Author feedback wizard. This is about to change, however: in the near future, such a tool will be launched in Scopus, which will allow institutions to provide automated feedback on their Scopus affiliation profiles. Prior to the launch of this tool, Scopus has already started to collect very detailed and structured feedback on its affiliation profiles from institutional leaders (such as university provosts and vice-chancellors), who need to sign off on any such profile improvements. Such

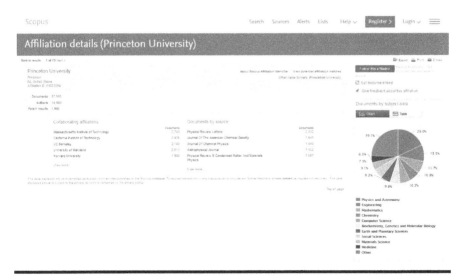

Figure 3.3   Scopus affiliation profile of Princeton University in the United States.

feedback includes the institute's preferred official name (in English) that should be displayed on the Scopus profile, the various legitimate ways of spelling that name, as well as the (potentially many) name variants; its primary address and various secondary addresses; the type of institute (e.g., a university or medical school) and institute's website; and finally, something which is a new feature in Scopus, whether there are any child organizations to the parent institute, such as a university hospital or a research institute. Such child organizations can be linked to the parent institute only if they already have a Scopus affiliation profile and ID of their own; the parent–child relationships between such profiles are shown as a hierarchical tree in Scopus.

By inviting such detailed feedback on Scopus affiliation profiles from the institutes themselves, especially on the many name variants that an institute may have, the quality of the profiles is expected to rise quite dramatically in the near future—primarily their average recall percentage. This is because adding all the possible name variants of an institute to the back-end data of its main profile—that is, the many different ways that the institute may be referred to in the affiliation field of the original publications—will ensure that the Scopus Affiliation Identifier algorithm will be able to group all incoming documents with those name variants into that institute's affiliation profile. That alone will increase the number of correctly classified documents in the profile, and thus its recall, significantly. Additionally, once the Institutional profile wizard has been launched, the average turnaround time before requested profile corrections will become visible on Scopus is expected to decrease substantially, to a matter of days eventually. And the new hierarchical structure of Scopus affiliation profiles, once fully rolled out, will also open up exciting new vistas for much more fine-grained analysis of institutional research performance, even down to the departmental level; the analysis capabilities of Elsevier's SciVal product, which is mostly built on Scopus affiliation profiles, will also benefit tremendously as a result.

3.6 Scopus Content Coverage

After having delved into each of the three basic units of Scopus, it is now time to turn our attention to one of the major pillars that Scopus is built on: the actual academic content that is indexed in Scopus. Being the largest curated abstract and citation database in the world today, and often selected by customers for the breadth and depth of its content, Scopus indexes three main types of scientific content: research journals, conference proceedings, and scholarly books. Currently in early 2017, this involves nearly 22,000 so-called *active* journals (which refers to the fact that Scopus indexes the latest issues of such journals when they are published, thus staying up-to-date), nearly 12,000 *inactive* journals (i.e., of which Scopus only indexes previous years or that have ceased publication), over 800 so-called trade journals of which 322 are active (with relevant content for specific industries and businesses), over 98,000 conference events totaling over 7.8 million conference papers, 558 active book series out of a total of over 1,200 series (together comprising more than 34,000 individual book volumes and over 1.3 million records), and more than 144,000 stand-alone books comprising another 1.3 million Scopus records as well. The decision to start indexing so many conference items (which took shape with the so-called Conference Expansion Program, finalized in 2014) was taken because conferences rather than journals are the main publication outlets for the fields engineering and computer science. Likewise, while book *series* were already indexed from the moment Scopus began, in 2013, a project was initiated to also add stand-alone books to Scopus (called the Scopus Book Titles Expansion Program), which are some of the main publication outlets in the arts and humanities and social sciences. The project was formally completed in 2015, but around 20,000 new books are still being added every year on an ongoing basis; this concerns monographs, edited volumes, major reference works, and graduate-level textbooks.

By adding books and conference proceedings to Scopus and thus substantially increasing its coverage of engineering, computer science, arts and humanities, and the social sciences, the citation counts for *all* indexed content in Scopus, including journal articles, become much more complete and accurate—after all, journal articles are referenced extensively in both academic book and conference content. This in itself enhances the author profiles and leads to more accurate *h*-indices for authors in those subject fields, as well as improved metrics in the affiliation profiles of institutes that specialize therein—which in turn allows them to better showcase their academic performance for things such as tenure promotions, institutional rankings, and national research assessments. But even without considering the newly added books and conferences, coverage across all subject fields in Scopus is already quite balanced to begin with: when looking only at active serial titles (i.e., journals, trade journals, and book series) within the four main subject categories, Scopus indexes over 4,500 titles in the life sciences, over 6,900 titles in the health sciences, over 7,500 titles in the physical sciences, and over 8,300 titles in the social sciences. This balance changes a bit when *all* document types and source titles are taken into account, so including conferences and one-off books, and when the relative proportion of subject fields is examined at the level of individual Scopus records rather than at the level of source titles. In Scopus, each source title is assigned to one or several of 27 subject categories, known as *All Science Journal Classification* (ASJC) codes, and these are further subdivided into a total of 334 subcategories. Hence, each Scopus record is assigned to the same ASJC code as the source title (i.e., journal, book, etc.) that the document has been published in. When these subject areas are compared across all 67 million Scopus records that are currently indexed, it is evident that medicine with 34% of all records is most strongly represented in Scopus as a single field, while the ASJC subject areas within physical sciences have the broadest coverage in Scopus overall—and also that the ASJC areas within social sciences are trailing somewhat (Figure 3.4). While this may seem

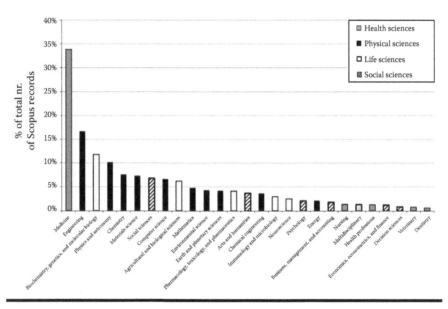

Figure 3.4 Percentage of Scopus records (across all document types and source titles) per subject area. Each source title in Scopus can be assigned to one or several of 27 ASJC codes; hence, the percentages above overlap across fields. ASJC codes themselves are categorized under one of four broad subject classifications: health sciences, physical sciences, life sciences, and social sciences.

surprising considering the balanced distribution of serial titles in Scopus within each of the four main categories, it does make sense when taking into account that researchers within, for example, medicine, biochemistry, or physics generally have a much higher publication output than someone publishing within, say, the history of art or accounting—so that a single biochemistry journal will have many more items indexed in Scopus than a single history journal.

Besides Scopus' broad coverage of subject areas (i.e., the breadth of its content), the source titles that are indexed are also generally span a large period of time and often go back to the first volume and issue published by a journal (an indication of the depth of the content). The very first publication year that can be found in Scopus is 1788, so even preceding the French Revolution, and concerns articles published by *Transactions of the Royal Society of Edinburgh*—while from 1798 onward, each subsequent year contains indexed items in Scopus. Not all content in Scopus, however, has its cited references indexed: when Scopus was launched in 2004, it was decided to add only cited references to documents indexed from 1996 onward; any earlier content did not have those references indexed at the time, which also restricts the citation counts and *h*-indices of authors who published the majority of their work in the 1970s and 1980s. That was one of the main reasons why the decision for yet another expansion project was taken, the *Cited References Expansion Program*.

This project, which started in late 2014 and was finalized in 2017, has added cited references to pre-1996 content going back to 1970, something which involved reprocessing of 4.5 million indexed items. In addition, around 7.5 million articles published before 1996 and that were not yet indexed were added to Scopus as well (including their references), involving the digital archives of more than 60 major publishers and often ensuring backfill to the first volume and issue of a journal in Scopus. In total, 195 million cited references were added to Scopus, giving these a major boost to a new total of more than 1.4 *billion* cited references indexed in Scopus overall. Since most of these 1.4 billion references represent citation links (insofar the referenced items are also indexed in Scopus), it is no surprise that this project has already led to a more than 41% increase in the average *h*-index of senior authors in Scopus who published before 1996, as an analysis performed at Elsevier has indicated. In addition, a recent study by Harzing & Alakangas (2016) found that in 2015, the average number of citations per academic (in their sample of 146 senior academics from the University of Melbourne) was around 100 citations higher for Scopus than for the Web of Science and also that Scopus provided more comprehensive coverage and a fairer comparison between disciplines than the Web of Science. And while the authors also considered the average number of citations for Google Scholar (GS), it is important to note that Scopus indexes scholarly content only from *selected* sources in its database, which has been subject to extensive quality control mechanisms. Scopus does not include gray literature and other non-authoritative sources.

If, after having been bombarded with all these numbers, this has not made a lasting impression yet, we could add to the mix that in addition to the source types mentioned earlier, Scopus also indexes over 38 million patents issued by the five main patent offices around the world: the U.S. Patent and Trademark Office (USPTO), European Patent Office (EPO), Japan Patent Office (JPO), World Intellectual Property Organization (WIPO), and UK Intellectual Property Office (IPO). These are sourced from Elsevier's sister company LexisNexis, and while they do not show much additional information in Scopus other than the patent title, authors, publication year, and issued patent number, they do provide useful links to full descriptions of them. Furthermore, it is worth highlighting the different document types that are indexed in Scopus: these include of course not only journal articles (both articles that describe original research and review articles), but also full conference papers, letters to the editor, notes, editorials, books, book chapters, and various miscellaneous items such as short surveys and errata—and more than 72% of indexed items have abstracts in Scopus. It is important to note that Scopus indexes every significant item from a journal; that is, all significant items that are selected by the editors and published in the journal issue are included in Scopus. There are also document types that are excluded from Scopus coverage, for instance, book reviews (as these do not describe original research) and conference abstracts (as these are often published again at a later time, e.g., as full conference papers or as journal articles that do get indexed in Scopus). Finally, with the increasing focus of governments and research funders around the world to make research results paid by taxpayers' money freely available to that same general public, it is good to know that the so-called Open Access status of indexed journals and book series is clearly indicated in Scopus (with the option of immediately linking to the full text). At the time of writing, Open Access status is indicated only on the level of serial titles, with the records of over 3,400 indexed journals and book series showing an *Open Access* indicator in Scopus.

All the different types of rich content that Scopus indexes and that have been summarized in this section originate from over 5,000 different academic publishers in total, including all the main publishers (Figure 3.5), such as Elsevier (11% of serial titles indexed), Springer Nature (9%), Wiley-Blackwell (5%), Taylor & Francis (5%), Sage (2%), and Wolters Kluwer Health (2% of titles)—but also many small publishers, including university presses and those from scientific societies. It is important to note that publishers still own the actual content displayed in Scopus—they deliver their content to the third-party suppliers of Scopus, usually in XML format (i.e., Extensible Markup Language) or otherwise as PDFs or (very occasionally still) in paper format. The suppliers then extract the relevant metadata and pass it on to the Scopus back-end, where extra information is added, the links between cited references and already indexed documents (i.e., the citation links) are established, and the author and affiliation profiles are updated or created; the entire throughput time between publisher delivery and appearance of a record on Scopus is around 5 days. Scopus is updated daily (unlike Scopus' nearest competitor for

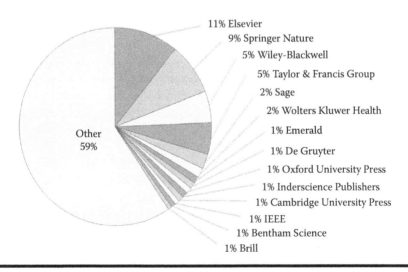

Figure 3.5 Distribution of academic publishers of serial titles that are indexed in Scopus. Percentages were calculated from the Scopus serial title list in October 2016.

instance, which is updated weekly), and every day, on average, up to 7,500 new Scopus records are added to the database. All publishers where this content originates from together represent 115 different countries from every corner of the globe, with content indexed from publications in more than 50 different languages, making Scopus the single *curated* abstract and citation database with both the best regional and the best global coverage.

3.7 Scopus Content Selection

While Scopus has the broadest and most comprehensive coverage of curated academic content, for any scientist, medical doctor, or university provost who uses Scopus data, it is equally critical to be able to trust that only the best-quality content gets indexed in Scopus, adhering to the highest scientific standards. Scopus has very stringent content selection procedures in place to make sure that is the case, and the team at Scopus has come a long way to establish the best possible policies and systems to support this process (e.g., Kähler, 2010, and many further improvements since then). The beating heart of it is formed by the Scopus Content Selection and Advisory Board (CSAB), an international group of scientists who are leading experts in each of the major scientific disciplines and who independently and with full authority review all serial titles for possible inclusion in Scopus. The CSAB currently consists of 17 different so-called Subject Chairs from all over the world, each reviewing the titles that were submitted within their own area of expertise, based on 14 different selection criteria in 5 categories—such as *academic contribution to the field* in the *content* category or *convincing editorial policy* in the *journal policy* category. Journals and other serial titles can be submitted for review to be included in Scopus via an online form on the Scopus information website, and such title suggestions need the full consent and cooperation of the publisher or editor of the title. Also, there are five minimum criteria that a serial title needs to meet before it is even eligible to enter the review process: it needs to publish content that is peer reviewed; publish on a regular basis, and have a registered ISSN; the articles should have English abstracts and article titles; have references in Roman script; and last but not least, the serial title should have a so-called *publication ethics and publication malpractice statement* that is publicly available. Submitted titles that meet these criteria enter a back-end system called the *Scopus Title Evaluation Platform* (STEP), and a dedicated content team at Scopus then collects a wealth of additional information about each title—such as electronic copies of the most recent issues, how often that title has already been cited by source titles that are indexed in Scopus, the international diversity of the authors and editors, and so forth—before it is released to the external CSAB for review. After their review, the relevant Subject Chair communicates their decision to the publisher or editor of the serial title and (in case that the title was not selected for Scopus indexing) may give suggestions for possible improvements.

3.8 Scopus Source Details Page, CiteScore, and Other Journal Metrics

Now that all the basics of Scopus have been covered, we can diverge into some exciting new developments and fruitful applications of the rich data indexed in Scopus. The aggregations of Scopus records on the level of both authors and affiliations were already discussed in previous sections, but this can just as easily be done on the level of source titles—which is especially useful for serial titles, such as journals, conference proceedings, and book series. Clicking on the name of a source title either in the Scopus search result page or from within a Scopus record will link to its Source details page, showing all the information that is available for that title, such as the publication years indexed in Scopus, the publisher, its print ISSN and/or electronic ISSN, the Scopus ASJC subject areas that have been assigned to it, and (if available) a link to its journal website (Figure 3.6). In December 2016, this page underwent a major overhaul when *CiteScore*™ metrics were launched, Scopus' brand new and freely available journal metrics (Colledge et. al., 2017; Van Noorden, 2016). With CiteScore, Scopus aims to provide a new, comprehensive, and transparent standard to help measure journal citation impact. Its basic calculation is very simple and can be verified by anyone: the number of documents in Scopus that cite the journal in question within a given year is divided by the total number of documents that were published by that journal in the three years preceding it. It is important to note that the numbers in both the numerator (i.e., the citing documents) and denominator (i.e., the published documents) involve *all* document types that are

Figure 3.6 Scopus Source details page for the journal *Nature*, showing how its CiteScore metric is calculated.

indexed in Scopus, except for AIPs, since those do not have indexed reference lists in Scopus (and thus do not provide citation links) and are only indexed in Scopus for a limited number of journals. If we take the renowned journal *Nature* as an example to illustrate the CiteScore calculation (Figure 3.6), we can see that its 2015 CiteScore of 14.38 is simply the number of 108,757 documents in Scopus that cited *Nature* articles in 2015 (the *Citation Count 2015*) divided by the 7,563 documents published by *Nature* in the period 2012 to 2014 *and* that are indexed in Scopus (the *Document Count 2012–2014*).

In addition to the yearly CiteScore, a static metric that is calculated based on an annual snapshot of Scopus data and is released in the second quarter of each year, the Scopus Source details page also provides the dynamic *CiteScore Tracker* (see Figure 3.6). This tracker is updated once a month and gives an indication of how the CiteScore is building up, either for the current year (if looked at after the annual CiteScore release) or for the previous year if that year's CiteScore is not yet available. Furthermore, since the absolute value of CiteScore itself can mean very different things in different subject areas (as citation patterns are known to vary wildly across disciplines), it is possible to compare CiteScore only across serial titles within the same subject area, as expressed by the *CiteScore Percentile* metric. Considering our example of *Nature*, which in Scopus has been assigned to the ASJC subject areas *medicine* and *multidisciplinary*, we see on the right side of Figure 3.6 that it is ranked as the 9th serial title out of a total of 1,549 medicine titles in Scopus—or in the 99th percentile of these titles, meaning that its citation performance is in the top 1% of medicine. And if from the drop-down menu, we were to select the category *multidisciplinary* instead, we would see that *Nature* is even ranked as the number 1 journal of the 98 Scopus titles in that category. The beauty of the CiteScore Percentile metric is that (unlike the absolute CiteScore value) this *can* be compared across titles in different subject areas. For example, if we take a look at the area *language and linguistics* (which is a subfield of the ASJC main field *arts and humanities*), we see that the number 1 performing journal in terms of its 2015 CiteScore is *Artificial Intelligence*; thus, even while this journal has an absolute CiteScore of only 5.41 (less than half of *Nature*'s 14.38), it can be considered as the *Nature* among Linguistics journals.

The Scopus source title page also shows two additional journal metrics that are themselves field-normalized, in the sense that (unlike CiteScore) their absolute values can be compared across serial titles in different subject areas. These are the *Source Normalized Impact per Paper* (or SNIP), which measures the number of actual citations received relative to the number expected for the serial title's subject area—and the *SCImago Journal Rank* (SJR), which measures *weighted* citations received by the serial title, where the weights assigned to citations depend on both the subject area and prestige (or SJR) of the citing serial title (Colledge et al., 2010). Both SNIP and SJR are complex metrics to calculate and cannot easily be validated in Scopus, so are less transparent and more difficult to interpret than CiteScore in that sense. Nevertheless, each available citation metric in Scopus has its own applications, depending on the context and specific questions to be addressed; each is useful in their own right by giving a unique perspective on the multifaceted diamond of an entity's research performance, and only by combining all these different views together does one gain a complete and truthful picture of it (Colledge & James, 2015). In conclusion, one of the greatest assets of CiteScore, SNIP, and SJR is that they are made available for free, as a service to the scientific community. Not only can these metrics be easily viewed and compared among serial titles at https://journalmetrics.scopus.com (with many useful filters), the Source details page of each serial title is now also accessible to non-Scopus subscribers as a free layer in https://www.scopus.com.

3.9 Scopus' Coming of Age: The Many Applications of Scopus Data

When Scopus was first launched as a search and discovery tool for researchers and librarians, few could have foreseen how the rich data that it contains would later come to power so many real-world applications beyond its original goals, especially in the area of research performance assessment. This section only intends to provide a bird's-eye view of these and offers a glimpse into what can be achieved with Scopus data.

- *Scopus analysis tools.* Within the Scopus web application itself at https://www.scopus.com, there are already many analysis tools available that enable users to easily visualize and track their search results, offering unique perspectives and insights into the underlying data structure that would be difficult to gain otherwise; just a few of the many examples include the journal analyzer (or *Compare sources*) tool, the *h*-index graph, and citation graphs and tables.
- *Scopus Custom Data.* For those customers who wish to integrate Scopus data into their own systems and tools, the option to purchase so-called SCD is available. SCD is an exact copy of either the entire online Scopus database at a fixed point in time or a subset thereof if only a limited set of publication years is requested—but as it represents only a snapshot of Scopus frozen in time, it is a static database (unlike Scopus itself, which is a dynamic database that is updated daily). SCD is delivered in XML format, just like the format of the Scopus web application itself, and has a very rich data structure with over 150 different XML elements. Some high-profile SCD customers include the Organization for Economic Co-operation and Development (OECD), the NSF, the ERC, Siemens, the Institute for Research Information and Quality Assurance (iFQ) in Germany, the Korea Institute of Science and Technology Information (KISTI), the Japan Science and Technology Agency (JST), Scimago Journal & Country Rank, Science-Metrix, and Sanofi.
- *Curriculum Vitae (CV) portals.* A related use of SCD that deserves separate mention concerns the use of sets of Scopus author profiles to populate

or enrich researcher CV portals with their publications and citations from Scopus. This may be especially useful for an institution (or even country) wishing to create a researcher ID profiling system for all of their researchers. Some examples of national researcher CV portals that have employed SCD are Currículo Lattes in Brazil, VIVO in the United States, and the Curriculum Vítae Normalizado (CVN) in Spain.

■ *Scopus APIs.* Where the use of SCD might be beyond the scope and available budget of most projects, a much simpler and cheaper solution could be to use Scopus APIs. These allow users to integrate content and data from Scopus into their own websites and applications (e.g., to show a live feed of citation counts from Scopus), with different APIs available on the level of basic Scopus records, author profiles, affiliation profiles, and serial titles. The use of basic Scopus APIs is free of charge to anyone, as long as Scopus' policies for using APIs and Scopus' data are honored—while Scopus subscribers have access to the full, more advanced APIs.

■ *Institutional repositories.* A special use case of Scopus APIs involves the building and enhancement of institutional repositories (IRs), which are systems to collect, preserve, and disseminate the intellectual output of an institution electronically; most universities have some sort of IR (also covered in Chapter 5 of this book). Many research institutions worldwide have built or enhanced their IRs using article metadata from Scopus, including high-profile ones such as Harvard University and the University of Queensland. While capturing all of an institution's research output into a single IR and keeping it updated can be a daunting task, this becomes much easier with Scopus data. Because of Scopus' comprehensive content coverage, as well as its automated profiles of virtually every institution and researcher in the world, a university's IR manager can simply query the relevant Scopus profiles (e.g., by using a Scopus API to periodically extract the metadata of new publications from them) to keep the IR complete and up-to-date.

■ *National research assessments.* To ensure that taxpayers' money is spent wisely and to promote research excellence, increasingly, governments evaluate their country's publicly funded universities through periodical research assessments. Some of the longest running and most sophisticated of these, such as the REF 2014 assessment in the United Kingdom and the ERA 2010, 2012, and 2015 exercises in Australia, have turned to Scopus as their exclusive citation data source to support them in their assessment, besides using expert peer review of research outputs—and also assessments in other countries, such as Portugal (the Fundação para a Ciência e a Tecnologia [FCT] exercise in 2014), Italy (Valutazione della Qualità della Ricerca [VQR] assessments and Abilitazione Scientifica Nazionale [ASN] accreditations in 2012, 2013, and 2016), and Japan (2016 assessment by the National Institution for Academic Degrees [NIAD] on the National University Corporation Evaluation Committee) have been supported by Scopus data (for details, see Schotten & el Aisati, 2014).

■ *Funding performance and research landscape studies.* Organizations funding science are keen to know the output of the research that they fund and the impact that it has both within academia and on society at large, thus justifying their investments and informing strategic decision-making. Scopus citation data are used for such analyses by some of the world's largest research funders, such as the NSF in the United States, which since 2016 has used Scopus data for its biennial *Science and Engineering Indicators* report, which is mandated by the U.S. Congress and recognized as the primary information source on U.S. research trends (Fenwick, 2016); as well as by the ERC, which has used Scopus data since 2011 to track the output and academic impact of research projects that it has funded.

■ *University rankings.* The rationale, use cases, and methodologies of university rankings, which have gained increasing popularity in recent years, are the topic of much of this book and are covered extensively in Section II. Some of the world's most prestigious rankings use Scopus data to power their analysis of the research and citation performance of universities; these include the *World University Rankings* (WUR) and various other rankings by THE (Fowler, 2014; see Chapter 8) and QS (Chapter 7), the *Best Chinese University Ranking* reports by ShanghaiRanking Consultancy (Chapter 6), the *Arab Region Ranking* by *U.S. News & World Report* (Chapter 12), as well as (although not strictly a university ranking) the *Scimago Journal & Country Rank* (Chapter 9; related to the SJR metric in Scopus); and the rankings by *Maclean's Magazine, Financial Times, Frankfurter Allgemeine Zeitung*, and the *Perspektywy University Ranking* in Poland. Also in 2017, a Scopus analysis of over 4,000 institutions will power the *National Institutional Ranking Framework* (NIRF) in India.

■ *Analytical reports.* Scopus data power a multitude of custom analytical reports, mostly commissioned by universities, R&D intensive corporations, funding bodies, government agencies, and international organizations and which are executed by Elsevier's Analytical Services team. Such analyses often shed light—for example, by using the Scopus author profile and affiliation profile histories—on the movements of researchers between institutes or countries (so-called brain circulation; Moed et al., 2013), sectors (e.g., between academia and industry), and research disciplines, as well as which types of collaborations prove to be particularly fruitful in terms of research impact. And when combined with other data, such as Scopus usage data, patent citations, or full-text download data from Elsevier's product ScienceDirect, even more powerful analyses become possible, offering unprecedented insights into things such as "knowledge transfer" between sectors. Some of the highest profile of these analytical reports include the 2017 report *Gender in the Global Research Landscape*; a 2014 report, in collaboration with the World Bank, that explored the development of research in sub-Saharan Africa; a 2015 report that explored six themes of the United Nations Sustainability Development Goals; a 2016 report highlighting the state of research on nanotechnologies; *World of Research 2015*, a 350-page book providing key statistics of the world's top 77 research nations; the 2015 report *America's Knowledge Economy: A State-by-State Review*; a 2013 report on the state of stem cell research; a 2014 report mapping the state of brain science and neuroscience research; and three reports commissioned by the UK Department for Business, Energy and Industrial Strategy (formerly known as Business, Innovation and Skills), comparing the international performance of the UK research base (2011, 2013, and 2017).

■ *How Scopus data power other Elsevier products.* Besides Scopus, Elsevier offers quite a few other tools and products that help researchers, research administrators, and funders in their respective workflows, all of which are well integrated with Scopus. The product that is perhaps most deeply integrated with Scopus is *SciVal*, which offers quick and easy access to the research performance of over 7,500 research institutions and 220 nations worldwide and which is built largely on Scopus data. Using advanced Data Analytics and High Performance Computing Cluster (HPCC) systems, SciVal allows its users to instantly process enormous amounts of data to generate powerful visualizations within seconds—for example, to get a quick overview of an entity's research performance, benchmark it with its peers, explore existing and potential collaboration opportunities, and analyze research trends to discover rising stars and hot new topics. Other Elsevier products that are linked well to Scopus include *Pure*, an advanced *Current Research Information System* (CRIS), that is, a research data capturing, analysis, and reporting tool used to centrally manage university administrations; *Mendeley*, a free and very popular reference manager and academic social network, as already covered in previous sections; and *ScienceDirect*, where more than 14 million full-text publications can be found of over 3,800 journals and 35,000 books published by Elsevier and its imprints and which shows for most documents the number of citations that they have received from Scopus records.

■ *Funding data.* A recent and exciting new development at Scopus involves the addition of a new layer of data to existing Scopus records and is known as the *Funding Data Expansion Project*. Funding information is captured from the full-text publications (usually present within the *Acknowledgments* section of a paper), and the funding body name, acronym, and grant number are then tagged with a so-called content enrichment framework tool. This funding information is available for all relevant Scopus records going forward, and backfill of funding data to 2008 will likely be completed by the end of 2017. This information will be extremely useful to funding organizations that wish to link research outputs to particular grants.

3.10 Imagine …

In this chapter, we have aimed to provide a broad, general overview of what Scopus is and the many things that can be accomplished with it, simplifying the life of researchers, students, health professionals, research managers, and funding agencies alike. We have also looked at the very diverse and broad-ranging content that is indexed in Scopus, including the stringent content selection policies that ensure that only the best quality gets indexed. But this chapter is by no means meant to be comprehensive and there remains room for much further study and self-learning of the many aspects of Scopus that are not covered here; Section 3.11 provides some excellent resources to get you started on that road. As our concluding remarks, we would like to remind ourselves that with Scopus, scientists have an outstanding tool at their disposal that can make their exciting research journey a more efficient, effective, and pleasant experience—but what really matters in the end, as with any great tool, is *how* we end up using it. Or, to put it in the words of Nobel Prize winner Albert Szent-Györgyi (who quoted freely after Schopenhauer): "Research is to see what everybody has seen and think what nobody has thought" (Schopenhauer, 1851; Szent-Györgyi, 1957). So while Scopus offers a bright and sharply focused flashlight (as well as a compass and Swiss army knife) to find our way around in the seemingly infinite world of research information out there, once such gems are discovered, these pieces of research information may themselves be combined together again in infinitely many and unexpected new ways to yield new insights, new ideas, and ultimately, new knowledge. But it is not knowledge per se that is the treasure waiting for us at the end of our long quest, but rather, how we choose to apply such newly gained knowledge to improve the quality of life for ourselves and for those we love around us, and for the betterment of society and humanity. And in order to get there, to return to Einstein's famous quote that we opened this chapter with, the only limit is the breadth of our imagination.

3.11 Resources and Further Reading

One of the best places to start looking for more detailed information on Scopus than what could be covered here is the Scopus information website at https://www.elsevier.com/solutions/scopus. This site also contains links to a very useful *Scopus Quick Reference Guide* and clear instruction videos (under the *Features* tab), as well as to the Scopus content coverage guide, Scopus title suggestion form, and Scopus serial and book title lists (under the *Content* tab). In addition, an excellent series of Scopus webinars on each of the topics discussed in this chapter can be found at https://blog.scopus.com/webinars. The Scopus blog at https://blog.scopus.com/ is in general a very good way to stay up-to-date on any new developments at Scopus.

Acknowledgments

The authors would like to express their deep gratitude and appreciation to all their colleagues at Elsevier throughout the years, especially those within Scopus Product Management, UCD, the Scopus development teams at Scopus Tech and Shared Platform, Research Products, Research Networks, Global E-Operations, Research Solutions Sales, Global Communications, and Marketing, as well as to all members, past and present, of the independent CSAB. Everyone mentioned here has, with their personal dedication, expertise, enthusiasm, and passion, played a pivotal role in building Scopus from the ground up and in ensuring its continued success to help researchers, doctors, and research managers to progress science and advance healthcare.

References

Allen, R. E., & Lidström, S. (2017). Life, the Universe, and everything—42 fundamental questions. *Physica Scripta, 92*(1), Art. no. 12501.

Boyle, F., & Sherman, D. (2005). Scopus™: The product and its development. *Serials Librarian, 49*(3), 147–153.

Colledge, L., de Moya Anegón, F., Guerrero Bote, V., López Illescas, C., el Aisati, M., & Moed, H. (2010). SJR and SNIP: Two new journal metrics in Elsevier's Scopus. *Serials, 23*(3), 215–221.

Colledge, L., & James, C. (2015). A "basket of metrics"—The best support for understanding journal merit. *European Science Editing, 41*(3), 61–65.

Colledge, L., James, C., Azoulay, N., Meester, W. J. N., & Plume, A. (2017). CiteScore metrics are suitable to address different situations—A case study. *European Science Editing, 43*(2), 27–31.

de Dijn, H. (1996). *Spinoza: The Way to Wisdom*. Purdue University Press, West Lafayette, IN.

de Heuvel, S. (2017). *Empowering Knowledge: The Story of Elsevier*. Elsevier, Amsterdam.

Fenwick, B. (2016). Tracking progress in U.S. science and engineering. *Elsevier Connect, 10 May*. Retrieved from https://www.elsevier.com/connect/tracking-progress-in-us-science-and-engineering

Fowler, N. (2014). Rankings provide a more complete picture of worldwide research. *Elsevier Connect, 19 Nov*. Retrieved from https://www.elsevier.com/connect/rankings-provide-a-more-complete-picture-of-worldwide-research

Harzing, A. W., & Alakangas, S. (2016). Google Scholar, Scopus and the Web of Science: A longitudinal and cross-disciplinary comparison. *Scientometrics, 106*(2), 787–804.

Hirsch, J. E. (2005). An index to quantify an individual's scientific research output. *Proceedings of the National Academy of Sciences of the United States of America, 102*(46), 16569–16572.

Jahn, R. G. (1996). Information, consciousness, and health. *Alternative Therapies in Health and Medicine, 2*(3), 32–38.

Kähler, O. (2010). Combining peer review and metrics to assess journals for inclusion in Scopus. *Learned Publishing, 23*(4), 336–346.

Moed, H. F., el Aisati, M., & Plume, A. (2013). Studying scientific migration in Scopus. *Scientometrics*, *94*(3), 929–942.

Schopenhauer, A. (1851). *Parerga und Paralipomena: Kleine Philosophische Schriften* (Vol. 2). A. W. Hayn, Berlin.

Schotten, M., & el Aisati, M. (2014). The rise of national research assessments—And the tools and data that make them work. *Elsevier Connect*, *17 Dec.* Retrieved from https://www.elsevier.com/connect/the-rise-of-national-research-assessments-and-the-tools-and-data-that-make-them-work

Szent-Györgyi, A. (1957). *Bioenergetics. Epigraph for Part II: Biological Structures and Functions.* Academic Press, New York.

Van Noorden, R. (2016). Impact factor gets heavyweight rival. *Nature*, *540*(15 December), 325–326.

Verlinde, E. (2011). On the origin of gravity and the laws of Newton. *Journal of High Energy Physics*, *2011*(4), Art. no. 029.

Verlinde, E. P. (2016). Emergent gravity and the dark universe. *arXiv*, 1611.02269. Retrieved from http://arxiv.org/abs/1611.02269

Wheeler, J. A. (1992). Recent thinking about the nature of the physical world: It from bit. *Annals of the New York Academy of Sciences*, *655*(1), 349–364.

Zijlstra, J., de Groot, S., el Aisati, M., Kalff, R., & Aalbersberg, Ij. J. (2006). Methods and systems for searching databases and displaying search results. U.S. Patent 20060112085.

Chapter 4

Google Scholar: The Big Data Bibliographic Tool

Emilio Delgado López-Cózar[1], Enrique Orduna-Malea[2], Alberto Martín-Martín[1], and Juan M. Ayllón[1]

[1]*Facultad de Comunicación y Documentación, Universidad de Granada, Colegio Máximo de Cartuja s/n, Granada, Spain*

[2]*Institute of Design and Manufacturing, Universitat Politècnica de València, Camino de Vera s/n, Valencia, Spain*

Contents

4.1 Introduction

Quantitative disciplines—such as Bibliometrics—are dependent to a great degree on their instruments of measurement. The more accurate the instrument, the better researchers will be able to observe specific phenomena. In the same way that the telescope fostered the evolution of astrophysics, improvements in bibliographic databases led to the advancement of Bibliometrics during the last decades of the twentieth century. The Internet (carrier), the web (contents), and search engines (content seekers) did all play a role to change the paradigm of bibliographic databases. The coming of academic search engines was the beginning of the era of robometrics (Jacsó, 2011), where web-based tools automatically index academic contents, regardless of their typology and language, providing data to third-party applications that automatically generate Bibliometric indicators. Google Scholar (GS) is the best robometric provider ever made. And GS never sleeps.

This chapter is devoted to the GS database. The main goal of this contribution is therefore to show how GS can be used for Bibliometric purposes and at the same time to introduce some Bibliometric tools that have been built using data from this source.

4.2 Birth and Development

GS is a freely accessible academic search engine (Ortega, 2014) which indexes scientific literature from a wide range of disciplines, document types, and languages, providing at the same time a set of supplementary services of great value. The fact that it displays the number of citations received by each document, regardless of their source, opened up the door to a new kind of Bibliometric analysis, revolutionizing the evaluation of academic performance, especially in the humanities and social sciences (Orduna-Malea et al., 2016).

However, facilitating Bibliometric analyses was never the main purpose of this platform. GS was conceived by two Google engineers (Anurag Acharya and Alex Verstak) who noticed that queries requesting academic-related material in Google shared similar patterns and that these patterns were related to the structure of academic documents. Consequently, these requested contents might be filtered, offering a new specialized service aimed at discovering and providing access to online academic contents worldwide (Van Noorden, 2014). Bibliometric analyses were never a goal; this was just a by-product brought on by its use of citations for ranking documents in searches.

Despite its simple interface and limited extra functionalities when compared to other traditional bibliographic databases, GS became rapidly known in the academic information search market after its launch in 2004. Both *Science* (Leslie, 2004) and *Nature* (Butler, 2004) reported the widespread use of this search engine among information professionals, scientists, and science policy makers.

The evolution of its website (2004–2016) can be observed in Figure 4.1 through the yearly screenshots captured from the Internet Archive's Wayback Machine.* An austere home page with a simple search box mimicking Google's general search engine and the *beta* declaration genuinely distinguished the first version. During these nearly 12 years since it was launched, the interface has barely changed. GS's improvements cannot be clearly perceived from looking at its interface, because it is in the engine itself where changes have taken place.

GS's ease of use, simplicity, and speed, as well as its multilingual and universal service and free of cost to the user, have contributed to its current popularity. To illustrate this, Figure 4.2 shows the worldwide search trends on Google for the main bibliographic databases (GS, PubMed, WoS, and Scopus). Since science policies might differ by country, we also offer data from trends in particular countries (United States, Belgium, Colombia, and India). Complementarily, Figure 4.3 shows the popularity of the search terms in a sample of regions and cities.

4.3 Characteristics of the GS Database

The fundamental pillars that sustain GS's engine are generally unknown not only to the final users, but also to journal editors (who tend to forget that online journals are—at the end of the day—web pages) and information professionals (who still often think in terms of the classic bibliographic databases). The consequences of these misconceptions might constitute total web invisibility for publications that are not represented on the web properly. Today, most students and researchers begin their searches of academic information in GS (Housewright et al., 2013; Bosman & Kramer, 2016). Thus, publications missing from GS's result pages may suffer significant losses in readership and maybe even a loss in citations as a result. This should be disquieting not only to journal editors and authors, but also to Bibliometricians.

---

* https://archive.org/web

Figure 4.1 Evolution of Google Scholar (November 2004–August 2016). (Courtesy of Internet Archive, San Francisco, CA.)

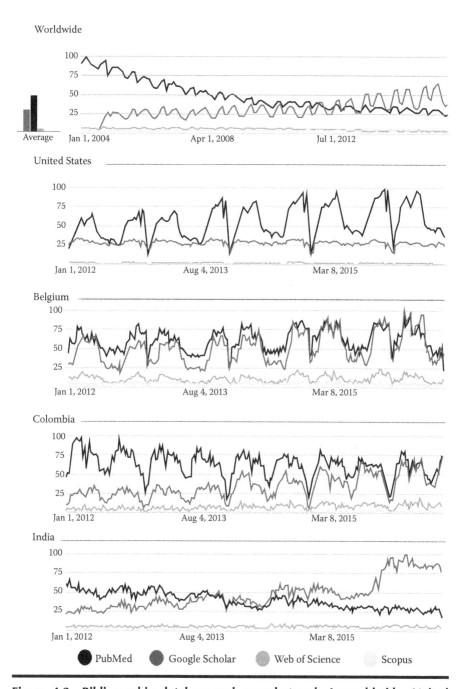

Figure 4.2 **Bibliographic database web search trends I: worldwide, United States, Belgium, Colombia, and India.** *Black*, PubMed; *dark gray*, GS; *medium gray*, WoS; *light gray*, Scopus. (Courtesy of Google Trends, Menlo Park, CA.)

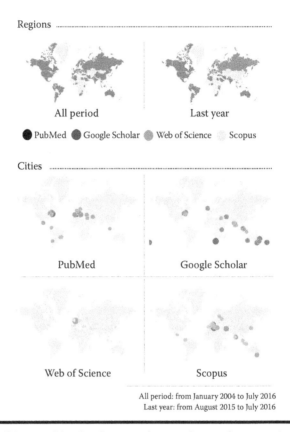

Figure 4.3   **Bibliographic database web search trends II: regions and cities.**
Black, **PubMed**; *dark gray*, **GS**; *medium gray*, **WoS**; *light gray*, **Scopus. (Courtesy of Google Trends, Menlo Park, CA.)**

This section describes how GS works as well as its main indexing requirements, with the hope that it will help users willing to extract—and contextualize—Bibliometric data from this database.

4.3.1 How Does GS Work? By Capturing the Academic Web

The approach of GS to document indexing clearly differs from that of classic bibliographic databases, which are based on the cumulative inclusion of selected sources based on their quality (mainly formal requirements). GS's approach, however, relies on the so-called academic web: any seemingly academic document available online will be indexed as long as a series of technical requirements are met.

Fortunately for print-only journals, GS indexes not only individual contributions available online, but also online catalogs and directories. Therefore, all those contributions that are not available to GS's crawlers (because there is not an online version, technical problems, or legal impediments such as paywalls) but indirectly catalogued in other bibliographic products (such as Dialnet) will also be indexed in GS.

GS operates in a similar fashion to Google's general search engine, managing a net of automated bots that crawl the web, looking for relevant information. These web crawlers are trained to identify academic resources, to extract their metadata and full texts (when available), and finally to create a bibliographic record to be included into GS's general index. If GS's crawlers are able to access the full text (either directly or indirectly by agreements with publishers), the system will also analyze the cited references in the document, and these references will be linked to the corresponding bibliographic records in GS as citations. The entire process is completely automated.

In short, and overlooking for now some exceptions that will be discussed in the following, GS only indexes *academic resources* deposited in the *academic web* that meet certain web requirements. These intellectual and technical formalities are detailed in the following.

1. Academic web

   The natural home for academic resources should be the academic web. This is GS's philosophy. Through the years, they have created a list of diverse institutions—both public and private—related in some way to academia, such as higher education institutions, national research councils, repositories, commercial publishers, journal-hosting services, bibliographic databases, and even other reputed academic search engines.

   Once indexed, these places are regularly visited by GS's spiders. Aside from these well-known academic entities, any natural or legal person is allowed to request their inclusion in GS's academic web space through the GS inclusion service.* Some of the accepted website types are DSpace, EPrints, other repositories, Open Journal Systems, other journal websites, and personal publications. Table 4.1 shows the number of records indexed in GS extracted from a small assortment of academic entities and services. However, these data should be considered rough estimations only. Table 4.1 provides a cursory view of the wide and diverse nature of the academic web space from which GS extracts information.

   While this method for finding out the number of records that GS has indexed from each domain has its own advantages, it also has some important shortcomings when performing Bibliometric analyses, among others:

   a. If data in Table 4.1 were collected again, results might be very different, maybe even lower! This is caused by the dynamic nature of the web (Lawrence & Giles, 1999). If a document becomes unavailable for any reason, GS will delete its presence from the database during one of its regular updates. GS's index reflects the web as it is at any given moment. Past is just past.

   b. The same document may be deposited in different places (journal website, IR, personal web page, etc.). Since the user experience would not be improved by

---

* https://partnerdash.google.com/partnerdash/d/scholarinclusions

Table 4.1 Showcase of Academic Entity Size in GS

Type	Entity	URL	Records	
			All	*2015*
Universities	Harvard University	harvard.edu	2,260,000	53,000
	National Autonomous University of Mexico	unam.mx	80,900	4,870
Research organizations	National Institute of Informatics	nii.ac.jp	12,900,000	279,000
	Max-Planck-Gesellschaft	mpg.de	105,000	4,240
Thematic repositories	arXiv	arxiv.org	402,000	53,200
	Social Science Research Network	ssrn.com	380,000	37,600
Publisher platforms	Elsevier	sciencedirect.com	8,750,000	483,000
	Nature Publishing Group	nature.com	449,000	30,500
Delivery services	Ingenta Connect	ingentaconnect.com	658,000	34,000
Databases	PubMed	ncbi.nlm.nih.gov	3,620,000	105,000
	Dialnet	dialnet.unirioja.es	2,830,000	101,000
Academic search engines	CiteSeerX	citeseerx.ist.psu.edu	1,020,000	13,200
	ResearchGate	researchgate.net	1,580,000	145,000

Note: Records obtained through the *site* search command, e.g., *site: harvard.edu.*

displaying the same document several times for the same query, GS groups together different versions of the same document (Verstak & Acharya, 2013). This process works fairly well for the most part, but it fails sometimes, mostly when the quality of the metadata is not very good, preventing the system from finding a match against its current document base for a new document that it is about to index, when a different version of that document has in fact been indexed before. Since each *unclustered* version may receive citations

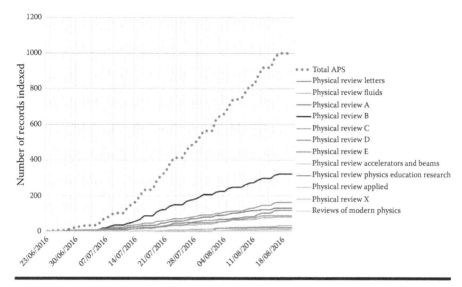

Figure 4.4   Indexation of APS journals in GS (August 2016).

independently, this problem affects any Bibliometric analyses that might want to be carried out using GS data, and so it is indispensable to group together all existing versions manually before carrying out any citation analyses.

c. The time elapsed since a resource becomes available online and GS's crawler's index it depends on the source. Harvard is Harvard, and indexing priorities do exist.

If we take a look at the last 1,000 articles published by the American Physical Society (APS)* indexed in GS (Figure 4.4), a *steplike* indexing, unrelated to the official publication periodicity, is observed. GS does not follow the classic and controlled issue-by-issue indexing process. Their technology made this practice obsolete, converting the academic web in a dynamic and uncontrolled web space. The irregularity and unpredictability of GS's indexing speed may bias some Bibliometric analyses if it is not taken under consideration.

2. Technical requirements

GS's inclusion policies also provide some guidelines for journal publishers and anyone who would like their contents to be correctly indexed in GS.† Some of these requirements are optional, whereas others are compulsory. Failure to comply with these rules may provoke an incorrect indexing or no indexing whatsoever for incompliant websites.

The first set of rules focuses on the websites. Their main objective is to ease content discovery. The website must not require users to install additional

---

* http://journals.aps.org
† https://scholar.google.com/intl/en/scholar/about.html

applications; to log in; to click additional buttons; or to use Flash, JavaScript, or form-based navigation to access the documents. In addition, the website should not display pop-ups, interstitial ads, or disclaimers.

The second set of rules is centered on the files that contain the full text. The system requires one URL per document (one intellectual work should not be divided into different files, while one URL should not contain independent works). Additionally, the size of the files must not exceed 5 MB. Although larger documents will be described in GS, their full text (including cited references) will be excluded. This may bias Bibliometric analyses since cited references included in doctoral theses and monographs (with files that tend to be larger than 5 MB) will be omitted.

HTML and PDF are the recommended file types. Additionally, PDFs must follow two important rules: (a) All PDF files must have searchable text. If these files consist of scanned images, the full texts (and cited references) will be excluded since GS's crawlers are unable to parse images. (b) All URLs pointing to PDF files must end with the *.pdf* file extension.

Last, the third set of rules encourages resource metadata description, establishing some recommendations on compulsory fields (title, authors, and publication date) and preferred metadata schemes (HighWire Press, EPrints, BE Press, and PRISM). Dublin Core may also be used as a last resort but is not encouraged since its schema does not contain separate fields for journal title, volume, issue, and page numbers.

If no metadata are readily available in the HTML meta tags of the page describing the article, GS will try to extract the metadata by parsing the full-text file itself. For this reason, GS also makes recommendations regarding the layout of the full texts: The title, authors, and abstract should be on the first page of the file (cover pages, used by some publishers, are strongly discouraged). The title should be the first content in the document, and no other text should be displayed with a larger font size. The authors should be listed below the title, with a smaller font size, but larger than the font size used for the normal text. At the end of the document, there should be a separate section called *References* or *Bibliography*, containing a list of numbered references.

4.3.2 Coverage of GS

Coverage and growth rate are essential aspects of any bibliographic database. This is not only about information transparency but also about context. Bibliometric analyses need to contextualize the results obtained since the database is just an instrument of measurement. Just like a chemist needs to check the calibration of his/her microscope to figure out the real dimensions of the observed elements in order to comprehend the underlying phenomena, an information scientist needs to verify which information sources are being indexed, the presence of languages, countries, journals, disciplines, and authors. Without context, Bibliometrics are just numbers. However, as the reader may probably imagine by reading the previous section, GS's coverage is, unfortunately, heterogeneous and still very much unknown.

Officially, GS indexes journal papers, conference papers, technical reports (or their drafts), doctoral and master's theses, preprints, postprints, academic books, abstracts, and *other scholarly literature* from all broad areas of research. Patents and case laws are also included. Content such as news or magazine articles, book reviews, and editorials is not appropriate for GS.* For example, any content successfully submitted to a repository will be included in GS regardless of its type. The lack of manual checking makes it impossible to filter documents by document type.

Moreover, the absence of a master list containing the publishers and sources that are officially covered has made many researchers wonder about its coverage. The continuous addition/removal of contents and sources as well as the technical exclusion of controlled sources makes the elaboration of any master list a chimera.

GS categorizes its documents into two independent collections: case laws and articles. The first group contains legal documents belonging to the Supreme and State courts of the United States. Since these documents are not used in Bibliometric analyses, we will not be studying them in this chapter. Regarding the collection of articles, we can distinguish the following contents:

1. Freely accessible online content

    This group includes all resources for which GS is able to find a freely accessible full-text link. If these documents include cited references, citing and cited documents will be automatically connected.

2. Subscription-based online content

    Most commercial publishers place the full texts of the articles that they publish behind paywalls. These documents are accessible only to people or institutions who have paid for the right to access them. By default, this would mean that GS might, at the most, have access to the basic bibliographic metadata for the articles (provided that the publisher makes the metadata available in the meta tags for each article and does not block GS's spiders by using robots.txt instructions), but probably not to the cited references, which are necessary to link citing and cited documents. Nowadays, however, GS has reached agreements with all the major publishers, and its spiders are able to collect all the necessary information from their websites (basic bibliographic metadata as well as cited references).

3. Content that is not available online

    When GS's spiders parse cited references inside a document, the system checks for matches for those documents in its document base in order to build a citing/cited relationship. If a match is not found for any of these

---

* https://scholar.google.com/intl/en/scholar/about.html

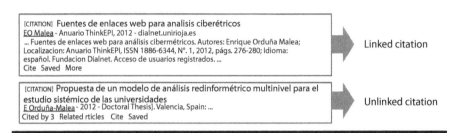

[CITATION] Fuentes de enlaces web para analisis ciberétricos
EO Malea - Anuario ThinkEPI, 2012 - dialnet.unirioja.es
... Fuentes de enlaces web para análisis cibermétricos. Autores: Enrique Orduña Malea;
Localizacion: Anuario ThinkEPI, ISSN 1886-6344, N°. 1, 2012, págs. 276-280; Idioma:
español. Fundacion Dialnet. Acceso de usuarios registrados. ...
Cite Saved More

Linked citation

[CITATION] Propuesta de un modelo de análisis redinformétrico multinivel para el
estudio sistémico de las universidades
E Orduña-Malea - 2012 - Doctoral Thesis]. Valencia, Spain: ...
Cited by 3 Related rticles Cite Saved

Unlinked citation

Figure 4.5   Linked and unlinked citations in GS.

references (because it does not exist or any variation in the bibliographic description prevents a correct match), it is added to the document base as a citation record (marked as [*CITATION*] when they are displayed as a search result). There are two types of citation records (Figure 4.5):

a.   Linked citations: Some bibliographic references found by GS in library catalogs and databases, with no full text available
b.   Unlinked citations: Bibliographic references found in the References section of a full text already crawled

4. Special collections

GS also indexes some collections from other Google services, such as Google Patents* and Google Book Search.† The inclusion guidelines are not very well documented in this regard, however.

GS officially states that the database automatically includes scholarly works from Google Book Search (excluding magazines, literature, essays, and the like). Additionally, any book cited by an indexed document will also be automatically included (as a [CITATION] record). In any case, users may also upload files directly to Google Books through their personal accounts. In a similar manner, the main reason to include patents is the fact that these resources are also cited in other documents indexed in GS. The fact that there are practically no patents in GS with zero citations reinforces this assumption.

4.3.3 Size of GS

The growth of GS is dynamic and irregular as academic sources become *GS-compliant*, new commercial agreements with publishers are attained, and old printed collections are digitized. This means that, apart from being continually indexing new materials as they are published, the retrospective growth of the platform is also remarkable. Figure 4.6 shows the number of records indexed in GS from 1700 to 2013 at two different times (May 2014 and August 2016). Although

---

* https://patents.google.com
† https://books.google.com

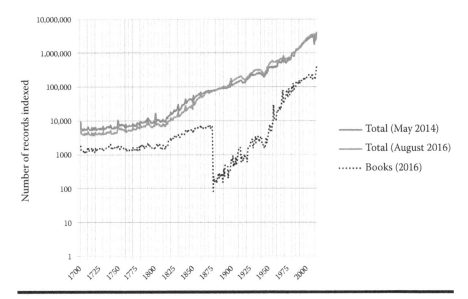

**Figure 4.6   GS evolution and retrospective growth. Note: The number of docu-
ments extracted after sending queries by using the *site* search command; books
from books.google.com. Patents are not included: patents.google.com and google
.com/patents do not work with the *site* command in GS.**

a logical correlation is obtained, differences are significant. For example, the difference between the two samples for the number of documents published in 2010 is of
more than 1,770,000 records!

GS is updated several times a week. In addition, the entire database goes
through a major update every 6–9 months, where it recrawls the academic web
and cleans data, making obsolete any study about its previous size. However, the
matter of its size constitutes a hot topic for the scientific community, fueled by the
scarcity of official details on this issue. There has been several attempts to unveil
the size of GS by using a variety of methods, all described in the scientific literature
(Jacsó, 2005; Aguillo, 2012; Khabsa & Giles, 2014; Ortega, 2014; Orduna-Malea
et al., 2015). Regarding the growth rate, Orduna-Malea et al. (2015), De Winter
et al. (2014), and Harzing (2014) proved that GS grows faster than other traditional
databases in all scientific fields.

However, even without considering the dynamic nature of the database, GS's
size cannot be calculated accurately due to several limitations of the search interface: the custom time range filter is not accurate, the *site* command is not exhaus-
tive, and the number of results per query is only a quick and rough estimation.

Savvy readers may be thinking of the possibility of performing several specific queries. However, no data export functionalities or application programming
interface (API) are available due to commercial constraints (probably some of the

stipulations in the agreements with the publishers discussed earlier). Last, only the first 1,000 results of each query can be displayed. Some techniques such as query splitting (Thelwall, 2008) may help but do not solve this limitation completely. Using web scraping in the result pages seems to be the only technical solution to perform big Bibliometric analyses (see Section 4.4.3), and even this approach has its own shortcomings. Moreover, doing this goes against GS's robots.txt, and they enforce this policy by blocking users who make too many queries too quickly.

4.4 Using GS for Bibliometric Purposes

While GS is a free, universal, and fast search engine with an impressive coverage, it lacks key search functionalities for information professionals (advance filtering, export features, sorting options, etc.). It is oriented to final users and content discovery, and it is not designed to work as a Bibliometric tool. These limitations may be what triggered the creation of applications which make use—either directly or indirectly—of its Bibliometric data. We can distinguish between the official products designed by the GS's team and other third-party products created by external and independent research teams.

4.4.1 Official Products

The GS team has designed two official products that make use of the bibliographic and Bibliometric data available in the core database. One of them is focused on authors (Google Scholar Citations [GSC]) and the other one on journals and the most cited articles in these journals (Google Scholar Metrics [GSM]).

4.4.1.1 Google Scholar Citations

Officially launched on November 16, 2011, this product lets users create an academic profile.* Users may build an academic résumé that includes all their contributions—provided that they are indexed in GS. The publications will be displayed in the profile, sorted decreasingly by number of citations received by default (they can also be sorted by year of publication and title). Users can search their publications by using their own author name (with all its variants) or by searching documents directly in order to add them to the profile, merge versions of the same document that GS has not automatically detected, and fix bibliographic errors. Most importantly, profiles are updated automatically as GS indexes new documents, with the possibility of asking the author for confirmation before making any changes to the profile.

---

* https://scholar.google.com/citations

In addition to the number of citations, the platform provides three author-level metrics: total number of citations received, *h*-index, and i10-index (number of articles which have received at least 10 citations), which are available for all documents (useful to senior researchers) and for the documents published in the last 5 years (useful to emerging scholars).

The platform also offers additional services such as personalized alerts, lists of coauthors, areas of interests, and list of authors by institution. By using this product, users can improve the dissemination and, potentially, the impact of their contributions. In short, authors can track the impact of their papers and other researchers' papers according to the data available in GS, as well as be constantly informed of new papers published by other authors. This makes GSC a very interesting, free, and easy-to-use research monitoring system.

The use of GS personal profiles can help to unveil much about an author's production and impact because, in a way, this product is a transition from an uncontrolled database to a structured system where authors, journals, organizations, and areas of interest go through manual filters (Ortega, 2014). However, some of the problems in GS are also present in this product.

4.4.1.2 Google Scholar Metrics

The Californian company surprised the Bibliometrics community again on April 1, 2012, by launching a journal ranking, commonly referred to as GSM.* Its characteristics and functionalities make GSM a unique and original product: Since its inception, GSM presented important differences compared to other journal rankings such as JCR and SJR (Cabezas-Clavijo & Delgado López-Cózar, 2012; Jacsó, 2012). GSM is a Bibliometric/bibliographic hybrid product, because in addition to Bibliometric indicators, it also displays the list of most cited documents in each publication. Moreover, its selection policies, coverage (journals, repositories, and conferences), architecture, and formal presentation are also different to other journal rankings.

Its coverage in its first edition (publications with at least 100 articles published during the 2007–2011 period and which had received at least one citation for those articles) and a categorization by language (for the last version available, 2011–2015 period, it covers the following languages: Chinese, Portuguese, Spanish, German, Russian, French, Japanese, Korean, Polish, Ukrainian, and Indonesian; Dutch and Italian are deprecated) were two of its most distinctive features. However, displaying journals sorted by their h5-index (*h*-index for articles published in a given 5-year period) instead of using a similar formula to the widely criticized JIF or the impenetrable SJR was probably its most distinctive feature. The platform also provided an internal search box which enabled users to locate journals not included in the general rankings, which were limited to the top 100 journals according to their

---

* https://scholar.google.com/metrics

h5-index. This search box presented the top 20 publications (again according to their h5-index) that matched the query terms entered by the user.

The lack of standardization, irreproducible data, or the amalgam of publication typologies available in the first version (Delgado López-Cózar & Robinson-García, 2012) made Cabezas-Clavijo & Delgado López-Cózar (2012) consider GSM as an immature product, although they acknowledged its potential as a source for evaluation of humanities and social sciences journals in the future. The GS team fixed some of the deficiencies mentioned in those early reviews and launched an improved version on November 15, 2012, introducing a subject classification scheme composed of eight broad categories and 313 subcategories. However, only journals published in English were classified in these categories, and only 20 publications were displayed in each of them (again the top 20 according to their h5-index).

Since then, the product has been updated every year. The last edition (July 2016) covers documents published in the 2011–2015 period. The total number of journals covered by this product is probably over 40,000 (Delgado López-Cózar & Cabezas-Clavijo, 2013). At any rate, various studies confirm that GSM covers more journals, published in more languages and countries, than JCR and SJR (Repiso & Delgado López-Cózar, 2013, 2014; Reina et al., 2013, 2014; Ayllón et al. 2016).

Despite its continuous improvements, Martín-Martín et al. (2014) did not hesitate in labeling GSM as a *low-cost* Bibliometric tool, with some powerful advantages (coverage, simplicity, free of cost) but some important shortcomings, most of them related to the difficulties of processing of journal data automatically, without any human intervention.

4.4.2 Third-Party Applications That Collect and Process GS Data

4.4.2.1 Harzing's Publish or Perish

If any third-party tool deserves a place in GS's Hall of Fame, this would undoubtedly be Publish or Perish (PoP). This free desktop application,* officially launched in 2006, lets users send queries to GS and GSC (also to Microsoft Academic) and collect all the available data, edit these, and obtain a set of Bibliometric indicators that can be exported outside the application, that is, all the features that many GS users would like to be able to carry out natively from the official platform and a few extras.

Despite its widespread use, the way that PoP works is relatively unknown. Some people think that it is an independent database, unrelated to GS, and some think that it makes use of some special API to access the information available in GS.

---

* http://www.harzing.com/pop.htm

None of that is true. PoP serves as a friendly interface between the user and GS (Harzing, 2013), but it is subject to the same limitations of any normal search in GS. The application transforms a user query and makes the appropriate request directly to GS's advance search. It then parses the results, displaying them on the PoP interface, at the same that it calculates additional metrics (total number of citations, authors per article, *h*-index, g-index, e-index, generalized *h*-index, AR-index, hlnorm, hl annual, multiauthored *h*-index, etc.).

Without any doubt, GS's coverage, together with PoP, has contributed to the democratization and popularization of citation analyses (Harzing & Van der Wal, 2008).

4.4.2.2 Scholarometer

This one is a less known but powerful tool developed by the School of Informatics and Computing at Indiana University–Bloomington, launched in 2009 (Kaur et al., 2012). It is a social tool which intends not only to facilitate citation analysis but also to facilitate social tagging of academic resources.*

Scholarometer is installed as an extension of the web browser and focuses primarily on the extraction of author data from GSC. Users can search authors through a search box or alternatively introduce the scholar's ID. In this last case, the system will extract and display the author's GSC profile, adding some extra functionalities and metrics, such as the article rank, and providing some data export functionalities as well, which are not available natively in GSC (except for one's own profile).

4.4.3 Third-Party Products That Make Use of GS Data

Last, we will introduce a set of Bibliometric products designed and developed by the Evaluación de la Ciencia y de la Comunicación Científica Research Group in Spain. The purpose of this is to illustrate the sort of products that can be generated using data from GS, GSC, and GSM.

4.4.3.1 H Index Scholar

H Index Scholar is a Bibliometric index which seeks to measure the academic performance of researchers from public Spanish universities in the areas of social sciences and humanities (SS&H) by counting the number of publications and citations received by their publications, according to data from GS (Delgado López-Cózar et al., 2014). Rankings are displayed broken down by four broad areas of knowledge (social sciences, humanities, law, and fine arts), grouped into 19 disciplines and 88 fields.† In each field, authors are sorted according to their *h*-index and

---

* http://scholarometer.indiana.edu
† http://hindexscholar.com

g-index, facilitating the identification of the most influential Spanish authors in the humanities and social sciences as of 2012.

This project (which covered more than 40,000 Spanish researchers) was the first serious attempt to study the suitability of GS for collecting the academic output of all SS&H researchers in a country.

4.4.3.2 Publishers Scholar Metrics

Book citation data are fundamental not only for researcher-level assessment in the SS&H, but also for getting an idea of the average impact made by each publisher. To date, there is no citation database that has anything resembling a comprehensive coverage of scientific books (the Book Citation Index is currently far from achieving this goal), and the evaluation of publishers has often been relied on reputational surveys.

Publishers Scholar Metrics* makes use of the GS database to construct a Bibliometric index to find evidence of the impact of book publishers, based on the citations received by the books published by the authors indexed in H Index Scholar (researchers from public Spanish universities, working in the SS&H, data collected in 2012). The main objectives of this product were to identify the core publishers by field as well as to demonstrate the suitability of GS data to carry out this task.

4.4.3.3 Proceedings Scholar Metrics

This product is a ranking of scientific meeting proceedings (conferences, workshops, etc.), in the areas of computer science, electric and electronic engineering, and communications, which have been indexed in GSM. To date, two editions (2009–2013 and 2010–2014) have been published (Martín-Martín et al., 2015a,b).

The main objective of this ranking is to compile an inventory of all conference proceedings included in GSM in these fields of knowledge, ranking them by their h5-index according to data from GSM.

4.4.3.4 Journal Scholar Metrics

Journal Scholar Metrics (JSM) is a Bibliometric tool that seeks to measure the performance of art, humanities, and social sciences journals by means of counting the number of citations that their articles have received, according to the data available in GSM.[†]

The main goals of JSM are, first, to calibrate the degree to which GSM covers international journals in the arts, humanities, and social sciences and, second, to identify the core journals in each of the disciplines (as well as other related journals

---

* http://www.publishers-scholarmetrics.info
† http://www.journal-scholar-metrics.infoec3.es

with thinner ties to the discipline) while offering a battery of citation-based metrics, complementing the indicators provided natively by GSM. By processing the metadata available in GSM, all journal self-citations (also called self-references) have been identified, allowing the calculation of the number of citations for each article excluding these journal self-citations and, consequently, the h5-index excluding journal self-citations and the journal self-citation rate.

4.4.3.5 Scholar Mirrors

This last product is a multifaceted platform which aims to quantify the academic impact of a whole scientific community (as a case study, we analyzed the community of researchers working on Scientometrics, Informetrics, Webometrics, and Altmetrics).*

The development of this product started with the identification of all the key members belonging to the studied community. To do this, those authors in the field who had created a public GSC profile were identified, and all the bibliographic information related to each of their contributions was collected.

Scholar Mirrors could be considered a deconstruction of traditional journal and author rankings, in alignment with the notion of multilevel analyses of a scientific discipline (documents, authors, publishers, and specific topics) instead of evaluating authors exclusively through the impact reported in the communication channels in which the research findings are published.

The platform offers a battery of author-level indicators extracted from a wide variety of academic social networks and profile services, making a total of 28 indicators (6 from GSC, 5 from RID, 9 from ResearchGate, 4 from Mendeley, 4 from Twitter). Additionally, authors are categorized as core (those authors whose scientific production is primarily done in the field of Bibliometrics) and related (those authors who have sporadically published Bibliometric studies, but their main research lines lie in other fields). The elements in the remaining levels of analysis (documents, journals, and publishers) are ranked according to the aggregated number of citations.

This multifaceted model facilitates the observation of the performance of the elements at different levels at the same time. It also displays in a clear way what each platform reflects about each of the authors, by way of their respective indicators, hence the mirror metaphor.

4.5 Final Remarks

Before presenting some conclusions on the potential Bibliometric uses of GS, we should stress an essential point which is already mentioned at the beginning of the

---

* http://www.scholar-mirrors.infoec3.es

chapter: the underlying reason that explains much of what has been discussed here. GS is, first and foremost, an academic search engine, a gateway to finding scientific information in the web. It was conceived with this sole purpose, and all its features and improvements are oriented to further this goal: connecting researchers with studies that may be useful to them. It was the members of the Bibliometric community who, upon becoming aware of the sheer wealth of scientific information available within this search engine, have repeatedly insisted on using this platform as a source of data for scientific evaluation. Once we acknowledge this truth, it is easy to understand the limitations of this platform for Bibliometric analyses.

At this point, curious readers may be asking themselves a fundamental question. Given the special nature of GS (unassisted, uncontrolled, unsupervised, lacking many advanced search features)—not to mention all its errors (authorship attribution, false citations, gaming, etc.), which is impossible to enunciate and describe all the details in a single book chapter—why use GS? The answer revolves around a concept: Big Data.

GS is undoubtedly the academic database with the widest coverage at this time, including journal articles, books, book chapters, dissertation theses, reports, proceedings, and so forth. No other scientific information system covers as many document types as GS, representing practically all formal and informal academic dissemination practices. Crawling the academic web allows GS to collect a great percentage of citations that are undetectable to other classic citation databases: GS is an academic Big Data system.

Classic citation databases constitute closed environments and are based on the controlled selection of a discrete number of journals: the elite. Classic citation indexes are built on citations among articles published in elite journals. This limitation—sometimes passed off as a feature—has its roots in Bradford's Law on literature concentration: few journals control a great percentage of science advancements. In practice, applying this law to select only a small portion of the scientific literature was primarily done because of technological and economic constraints which have now disappeared for the most part.

When GS started collecting data about all seemingly academic documents deposited in trusted web domains (but which had not necessarily passed any other external controls, such as peer review), they broke with those, until then, common selection practices, to which researchers (as well as journal editors) had gotten used to. GS did not limit itself to the scientific world in the strict sense of the term, but instead embraced the whole academic world.

Although important errors do exist, Big Data transform them in inherent aspects of the database. Even with "dirty" data, it is able to distinguish the wheat from the chaff. However, citations mixed regardless of the source represent the real fraught between apocalyptic and integrated. Citations among any academic resource shape the GS's author credential. The excellence view is shortsighted.

Citation-based author performance evaluation through traditional bibliographic databases (WoS or Scopus) for researchers in the humanities and social sciences makes no sense, as these platforms do not cover the main venues where these authors disseminate their research results. The launch of GS back in 2004 meant a revolution not only in the scientific information search market but also for research evaluation processes, especially for disciplines where results are not usually published as articles in journals published in English. In order to cover the need to evaluate those disciplines, several applications and products that make use of GS data (with a much better coverage of the research outputs in those disciplines) have been developed. GS presents not only many challenges, but also a lot of potential as a source of data for Bibliometric analyses.

References

Aguillo, I.F. (2012). Is Google Scholar useful for bibliometrics? A webometric analysis. *Scientometrics*, 91(2), 343–351.

Ayllón Millán, J.M., Martín-Martín, A., Orduna-Malea, E., Delgado López-Cózar, E. (2016). Índice H de las revistas científicas españolas según Google Scholar Metrics (2011–2015). EC3 Reports, 17.

Bosman, J., & Kramer, B. (2016). Innovations in scholarly communication—Data of the global 2015–2016 survey. Zenodo. https://zenodo.org/record/49583#.WUQO5WjyjIU.

Butler, D. (2004). Science searches shift up a gear as Google starts Scholar engine. *Nature*, 432(7016), 423.

Cabezas-Clavijo, A., & Delgado López-Cózar, E. (2012). Scholar Metrics: The impact of journals according to Google, just an amusement or a valid scientific tool? EC3 Working Papers, 1.

Delgado López-Cózar, E., & Cabezas-Clavijo, Á. (2013). Ranking journals: Could GSM be an alternative to Journal Citation Reports and Scimago journal rank? *Learned Publishing*, 26(2), 101–113.

Delgado López-Cózar, E., Orduña-Malea, E., Jiménez-Contreras, E., Ruiz-Pérez, R. (2014). H Index Scholar: El índice h de los profesores de las universidades públicas españolas en humanidades y ciencias sociales. *El profesional de la información*, 23(1), 87–94.

Delgado López-Cózar, E., & Robinson-García, N. (2012). Repositories in Google Scholar Metrics or what is this document type doing in a place as such? *Cybermetrics*, 16(1).

De Winter, J.C.F., Zadpoor, A.A., Dodou, D. (2014). The expansion of Google Scholar versus Web of Science: A longitudinal study. *Scientometrics*, 98(2), 1547–1565.

Harzing, A.-W. (2013). *The Publish or Perish Book*. Melbourne: Tarma Software Research.

Harzing, A.-W. (2014). A longitudinal study of Google Scholar coverage between 2012 and 2013. *Scientometrics*, 98(1), 565–575.

Harzing, A.-W., & Van der Wal, R. (2008). Google Scholar as a new source for citation analysis. *Ethics in Science and Environmental Politics*, 8(1), 61–73.

Housewright, R., Schonfeld, R.C., Wulfson, K. (2013). Ithaka S+R, Jisc, RLUK UK Survey of Academics 2012. Ann Arbor, MI: Inter-university Consortium for Political and Social Research. http://www.sr.ithaka.org/publications/ithaka-sr-jisc-rluk-uk-survey-of-academics-2012/?

Jacsó, P. (2005). As we may search—Comparison of major features of the Web of Science, Scopus, and Google Scholar. *Current Science*, 89(9), 1537–1547.

Jacsó, P. (2011). Google Scholar duped and deduped—The aura of "robometrics." *Online Information Review*, 35(1), 154–160.

Jacsó, P. (2012). Google Scholar Metrics for Publications: The software and content features of a new open access bibliometric service. *Online Information Review*, 36(4), 604–619.

Kaur, J., Hoang, D.T., Sun, X., Possamai, L., JafariAsbagh, M., Patil, S., Menczer, F. (2012). Scholarometer: A social framework for analyzing impact across disciplines. *PloS One*, 7(9), e43235.

Khabsa, M., & Giles, C.L. (2014). The number of scholarly documents on the public web. *PloS One*, 9(5), e93949.

Lawrence, S., & Giles, C.L. (1999). Accessibility of information on the web. *Nature*, 400(6740), 107–107.

Leslie, M.A. (2004). A Google for academia. *Science*, 306(5702), 1661–1663.

Martín-Martín, A., Ayllón, J.M., Orduna-Malea, E., Delgado López-Cózar, E. (2014). Google Scholar Metrics 2014: A low cost bibliometric tool. EC3 Working Papers, 17.

Martín-Martín, A., Ayllón, J.M., Orduna-Malea, E., Delgado López-Cózar, E. (2015a). Proceedings Scholar Metrics: H Index of Proceedings on Computer Science, Electrical & Electronic Engineering, and Communications according to Google Scholar Metrics (2010–2014), EC3 Reports, 15.

Martín-Martín, A., Orduna-Malea, E., Ayllon, J.M., Delgado Lopez-Cozar, E. (2015b). Proceedings Scholar Metrics: H Index of Proceedings on Computer Science, Electrical & Electronic Engineering, and Communications according to Google Scholar Metrics (2009–2013), EC3 Reports, 12. Available at: http://arxiv.org/abs/1412.7633.

Orduna-Malea, E., Ayllón, J.M., Martín-Martín, A., Delgado López-Cózar, E. (2015). Methods for estimating the size of Google Scholar. *Scientometrics*, 104(3), 931–949.

Orduna-Malea, E., Martín-Martín, A., Ayllón, J.M., Delgado López-Cózar, E. (2016). *La revolución Google Scholar: destapando la caja de Pandora académica*. Granada, Spain: UNE.

Ortega, J.L. (2014). *Academic Search Engines: A Quantitative Outlook*. Oxford: Elsevier.

Reina, L.M., Repiso, R., Delgado López-Cózar, E. (2013). H Index of scientific Nursing journals according to Google Scholar Metrics (2007–2011). EC3 Reports, 5.

Reina, L.M., Repiso, R., Delgado López-Cózar, E. (2014). H Index of scientific Nursing journals according to Google Scholar Metrics (2008–2012). EC3 Reports, 8.

Repiso, R., & Delgado López-Cózar, E. (2013). H Index Communication Journals according to Google Scholar Metrics (2008–2012). EC3 Reports, 6.

Repiso, R., & Delgado López-Cózar, E. (2014). H Index Communication Journals according to Google Scholar Metrics (2009–2013). EC3 Reports, 10.

Thelwall, M. (2008). Extracting accurate and complete results from search engines: Case study Windows Live. *JASIST*, 59(1), 38–50.

Van Noorden, R. (2014). Google Scholar Pioneer on search engine's future. Nature News.

Verstak, A., & Acharya, A. (2013). Identifying multiple versions of documents. Washington, DC: U.S. Patent and Trademark Office.

Chapter 5

Institutional Repositories

Maria-Soledad Ramírez-Montoya
and Héctor G. Ceballos
Tecnologico de Monterrey, Monterrey, Mexico

Contents

5.1 Introduction

In the context of the open access (OA) movement, repositories provide a window for scientific and academic communities to preserve and share knowledge. Glassman & Kang (2016) stressed the importance of educational institutions developing technologies that support an open educational system. Accordingly, Ramírez (2015) mentioned that OA is an opportunity to improve the transfer and dissemination of knowledge and discuss the relevance of repositories as a space to accommodate and recover scientific and educational production. In this sense, institutional repositories (IRs) not only constitute a support for the open dissemination of knowledge, but also bring with them challenges that must be worked out in organizational cultures.

A repository is a technological platform of OA to knowledge, which is directed to the storage, preservation, and diffusion of the production generated in the institutions. MacIntyre & Jones (2016) emphasized the support of IRs for research and the visibility that they can bestow on institutions. Fontes Ferreira & Souza de Silva (2015) emphasized the idea that the more academic tools people have, with the support of repositories, the better academic level they will reach, as well as a modern way to study. Empirical studies have provided evidence of the positive role of repositories; for example, Koler-Povh et al. (2014) presented a study on the IR of the University of Ljubljana in early 2011,* where they found that 89% of its visitors came from other institutions. This shows how important IRs are to publish information and how communication between institutions and universities con-tribute to open education.

In contrast to the advantages provided by IRs, it is also necessary to state the challenges existing in the environment of the open education movement. Davis et al. (2010) presented the problem of the poor response to open educational models, despite the great investment in infrastructure that is taking place. Cragin et al. (2010) analyzed the cognitive processes that lead authors to share or not share their work, based on the study of their cognitive processes and the *open sharing* culture. Another challenge lies in the registration of resources so that knowledge can be *discovered* and used; for example, González (2016) found in his study that meta-information in articles of IRs may be incomplete and that could generate difficulties to access this information, hence the importance that authors and library staff should consider methods to improve the indexing of articles in order to be discovered.

The motivations for institutions to have a repository can be very varied. Being a set of centralized web services, the objective of an IR is to organize, manage, pre-serve, and disseminate digital materials, mainly scientific and academic production created by an institution and its members. In this way, IRs provide institutions with the possibility of improving their position in rankings, ensuring the preservation of their organizational memory, gaining visibility and presence on the web, increas-ing the impact (citation) of authors, encouraging scientific collaboration (interna-tionalization), supporting innovation (research projects), and recently being able to respond to national policies of open sharing of the production generated in publicly financed projects.

The OA movement in the previous decade and IRs developed by universities and academic libraries as part of this movement have challenged the traditional model of the school communication system. Researchers such as Cullen & Chawner (2011) examined the growth of IRs alongside the OA movement and reported the findings of a national survey of academics highlighting the conflict between the principles and rewards of the traditional school communication system and the benefits of the OA movement.

In different latitudes, the motivation to have IRs is evident. Figure 5.1 shows the distribution by country of the 2,824 IRs listed by February 7, 2017, in the Directory of Open Access Repositories (OpenDOAR).* As can be seen, United States

---

* http://www.opendoar.org/

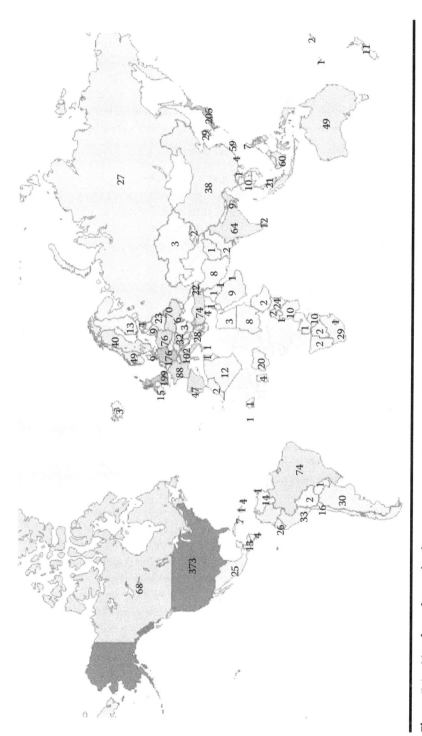

Figure 5.1 Number of repositories per country. (Courtesy of OpenDOAR, http://www.opendoar.org/, February 7, 2017.)

is the country with the most repositories listed (373); nevertheless, Europe together has more than three times this number of repositories.

Scientific and academic production that is of interest for institutions is stored in repositories. Scientific productions such as papers, articles, chapters, books, journals, and datasets are highly valued documents to share in the research community. Also, academic productions such as cases, learning objects, videos, notes, and presentations are useful for the teaching and student community. This production must be cataloged in such a way that it can be found to be used. Shukair et al. (2013) mentioned that in order to facilitate the discovery, access, and reuse of resources, semantic interoperability must be considered, that is, a thesaurus of concepts that describe the information and a new model of metadata to identify the types of reusable resources, which facilitate their discovery and ensure a minimum consistency that is directly related to the definition of a conceptual architecture for a federated resource deposit.

5.2 Open Access

The OA movement aims at the online availability of scientific products free of cost and restrictions on the use of the information contained within. In the dissemination of scientific production, two main routes for OA can be observed: to publish in an OA journal (Gold OA) or to archive copies of the article in a repository or website (Green OA). Björk et al. (2013) presented information about what Green OA is and its differences with Gold OA and analyzed the legal considerations of the publications and the ease of its administration. They concluded that the Green OA technical infrastructure is becoming more robust due to the increase in the number of IRs. Also, Johnson et al. (2015) mentioned the differences between the two routes, which have different purposes, and also agreed that the interest in Green OA is greater, since it is focused on IRs of universities. In addition, they mentioned aspects of the costs and of how each of them relates in the society.

Digital preservation and copyright control in IRs must be strengthened to increase participation. Kim (2011), through a study, demonstrated the motivation given to faculty staff to adopt Green OA, where the faculties were motivated by both preservation and copyright. Teachers contributed to IRs to make their material extensively accessible and not lose the benefits of OA. Another very positive motivation for authors is the possibility of citation that gives them the open publication. Gargouri et al. (2010) demonstrated that articles written through Green OA are cited more frequently than articles requiring subscription. The previously mentioned is due to the quality advantage of these items as users have more freedom to choose which articles to cite.

As we see benefits, there are also challenges faced by Green OA. Antonczak et al. (2016) claimed that there are obstacles to creating repositories. The people who produce and create these repositories are creative, but they could create such repositories for purposes other than to benefit society. On the other hand, Wallace (2011) presented the development of Publishing and the Ecology of European Research (PEER), in which they elaborated an observatory with more than 44,000 manuscripts in Green OA. One of the challenges of this development was the difference in manuscript types and metadata formats. The author focuses on the Green OA based on a study in the United Kingdom, which also compares the attitudes and behaviors of OA authors and users and the challenges of developing PEER's Green OA. Another challenge lies in the number of possibilities for open publishing.

Studies related to publication in repositories have also been carried out. Dawson & Yang (2016) did a study of the Registry of OA Repositories database to find trends in how major IRs handle copyright and OA. The authors found that there is no clear way for scientific publications to achieve OA status and suggest that if you want to start an IR program, libraries need to be involved with the faculty to help them obtain copyright permits. It is necessary to educate faculty staff about copyright, as this is one of the main barriers to the growth of repositories. It is recommended that librarians become experts on the subject of copyright to assist in the process.

5.3 Rankings

The evaluation of repositories has generated the possibility of grouping them according to their characteristics. One of the most recognized rankings is Webometrics, which arose in the Superior Center of Scientific Investigation. Aguillo et al. (2010) presented Webometrics and evaluated the repositories according to the following characteristics:

1. Listed in OpenDOAR
2. Being institutional or thematic
3. Contains scientific articles
4. Included in the domain of the institution
5. Each file measured according to size in number of pages; should be a PDF file
6. Must be found in Google Scholar (GS) and must have visibility in external links

Another ranking focuses on success characteristics of repositories oriented to scientific work. Marcial & Hemminger (2010) studied a sample of 100 scientific data repositories (SDRs) and generated different segments or classifications of them. The characteristics of the SDRs were explored to identify their role in determining groups and their relation to the success of the group: if it received funding support, if the support came from different sponsors, the size of the SDR, and if there was a preservation policy.

5.4 Platforms and Protocols

Open platforms are multiple, and this has increased the possibilities for repositories. Pinfield et al. (2014) provided an analysis of different platforms from 2005 to 2012 in different parts of the world where they reviewed the growth of OA repositories, using data from the OpenDOAR project, as well as the growth of type proposals in each country. It shows the map of IRs, their development, and operation. The research focused mainly on North America, Western Europe, and Australasia. Since 2010, there has been growth in repositories in East Asia, South America, and Eastern Europe. Globally, repositories are predominantly institutional and multidisciplinary and in English. Typically, they use OA software, but they have some problems with licensing. Tzoc (2016), on the other hand, carried out a study to see the platforms of IRs most used in universities in the United States. The results of the 67 institutions taken into account for the study were DSpace, CONTENTdm, Islandora, and North Carolina Digital Online Collection of Knowledge and Scholarship.

Figure 5.2 shows the main platforms used by the 2,824 IRs listed in OpenDOAR by February 7, 2017. As can be seen, DSpace is the platform mainly used (48%), followed by EPrints (14%), and Digital Commons (5%); meanwhile, 27% of the remaining repositories use diverse platforms.

The Open Archives Initiative Protocol for Metadata Harvesting (OAI-PMH) provides an automatic mechanism for transferring information between repositories.* Through this protocol, repositories become *data providers* by exposing structured metadata describing resources arranged in collections. Other repositories, called *service providers*, use OAI-PMH service requests to harvest that metadata.

Figure 5.3 shows the percentage of repositories supporting the protocol OAI-PMH (bars) and the percentage of resources hosted in those repositories, as listed in OpenDOAR by February 7, 2017. As can be seen, despite 70% of IRs supporting this protocol, less than 45% of the resources can be accessed through it.

Metadata vocabularies used for describing resources are a key aspect for linking and visibility of repositories. Arlitsch & O'Brien (2012) sought to prove a theory that transforming metadata schemas of IRs will lead to an increase in GS indexing. The report indicates that repositories using GS's recommended data schemas and those expressed in HTML *meta tags* have a much better indexing rate. The ease with which search engine crawlers can navigate a repository affects the indexing rate as well. The importance of this research is that the lack of visibility in GS will limit the visibility of the resources stored in these repositories, preventing a more important role in the increase on the number of citations.

---

* https://www.openarchives.org/pmh/

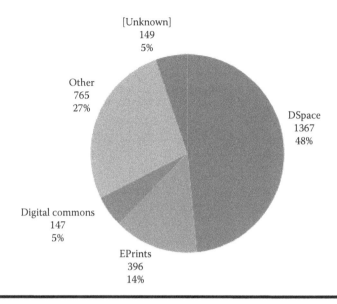

Figure 5.2 IRs by platform. (Courtesy of OpenDOAR, http://www.opendoar.org/, February 7, 2017.)

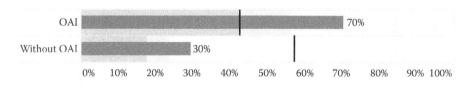

Figure 5.3 Repositories supporting the protocol OAI-PMH. (Courtesy of OpenDOAR, http://www.opendoar.org/, February 7, 2017.)

5.5 Challenges for IRs

Limitations of platforms and protocols for repositories are present in different stages. Way (2010) showed a problem of archiving articles in repositories; he conducted a study using GS to examine the OA availability of the library and information science (LIS) and examined whether GS was able to find the links to the full texts in case that there were OA versions of articles. It was found that article archiving is not commonly practiced, but GS is an effective tool for finding LIS articles. Laakso & Björk (2012) described how different methods and applications have appeared over the years so that repositories can be used in different ways and purposes. The issue of costs is also a limiting factor. Burns et al. (2013) warned that little is known about the costs incurred by an academic library when implementing and managing IRs and the value that these repositories offer to their communities.

Beyond technological issues, human factors limit the deposit and sharing of materials. Wu (2015) estimated that only 15–30% of academics deposit their work in IRs. These studies suggest that the most common reasons that contribute to mistrust are plagiarism and copyright issues, the impact of their citations, credibility, and the time required to enter objects into the repository. The problems with most impact in small institutions are the limited budget, lack of planning, and lack of technological expertise. To promote the development of its repository, the university that publishes this paper proposes to use more of its IR materials in the classroom, to start producing audiovisual content and to integrate IRs with classroom management systems. In addition, it is proposed to use the repositories as a space to apply what was learned in class, publishing journals made completely based on the work of the students; this can integrate many disciplines such as graphic design and research.

Several analyses have been conducted on IRs, both technical and on usability. For example, in the perspective of users of IRs, Jean et al. (2009) presented an investigation about the use that users give to IRs and their perception about them. In their research questions, one can appreciate a concern about how frequently users access the repository, for what purpose they do it, whether or not they use the content that they find, and if they recommend its use to their colleagues. Berquist (2015) explained how repositories have grown and the implications that they have for journals, researchers, universities, and so on. It outlines the problems that arise from the IRs and calls for solutions for authors to work better with institutions. Berquist (2016) focused on how to improve the ways of working and minimizing the time and effort required for the deposit of articles in repositories.

Other studies have also been able to add value to the growth of IRs. Schöpfel (2013) proposed five ways to add value to IRs:

- *Quality*: This can be achieved by constraining the acceptance of any type of work, giving unrestricted access to the full text, and having validation of the institution.
- *Metadata*: Using standardized schemas, you can enrich the description of content that is not explicitly written in the text. This increases the likelihood of finding resources in repositories.
- *Format*: IRs must contain the complete text, not only metadata, independently of having files with formats such as MS Word and PDF. For scanned files, it is important to use optical character recognition (OCR) before indexing them.
- *Interoperability*: It is important that repositories connect to each other. For this, it is necessary to have a standard format for describing resources, policies for the exchange of information, etc.
- *Services*: You must have an advanced search, display, download, and navigation system to improve the user experience.

In addition, it is suggested to use another type of nontextual format and enrich the resources with *deep access*, linking them with other information, alternative representations, and activities.

Open technologies and the possibilities that they provide will have a predominant role in the next generation of repositories. Sotiriou et al. (2016) presented research on education using Open Discovery Spaces, an initiative that premodernizes schools with a large implementation of open-scale methodologies using technological innovation. The researchers applied the model to different schools and observed how student connections and overall satisfaction grew, as technological maturity increased considerably and as those who participated gradually shared their knowledge. Rodés-Paragarino et al. (2016) mentioned several studies on the use and adoption of repositories and other digital educational resources, making clear that there is a challenge to strengthen those that can still give repositories in terms of connection.

In the open educational movement, repositories still have a long way to go. Xia et al. (2012) mentioned that, although many policies have proven to be effective and positive, some others have no impact on the development of repositories. Thus the enactment of policies of open education movement, by itself, will not change the existing practices of academic self-archiving. A clear aspect was evidenced by Ezema (2011), who presented the necessary elements for the development of the open educational movement and the use of repositories in Nigeria. The article proposes the development of a new technological culture, awareness, financing of universities, development of information and communication technology infrastructure, and the presentation of theses and electronic dissertations.

5.6 Conclusions

Despite the growth of IRs reported by OpenDOAR being linear, the number of records that they hold increases exponentially. By February 7, 2017, the 2,824 IRs listed in OpenDOAR accounted for 105.75 million records, with an average of 38,300 records per repository.

As described earlier, this growth is motivated both by the support given by universities to the OA movement and by the relevance of web presence in university rankings. And while the metadata format and harvest protocols provide facilities for inspecting the contents of repositories, finding the right resources for research or teaching materials remains a difficult problem to solve. Resources are cataloged under diverse classification systems, and despite efforts such as ORCID,* there are no unique identifiers of authors associated with all available materials. Recognizing equivalent resources, authors, and affiliations across repositories is another hard task.

---

* https://orcid.org/

As can be seen, data analytics plays an important role in this scenario. The question remains open whether the impact (citation) of scientific publications published in OA through IRs is greater than that of restricted access publications and whether this is independent of the quality of the journal or the paper. Identifying related resources and determining the characteristics of the most downloaded resources are other areas of application.

Finally, although IRs represent a support tool for institutions, there is still much work to be done in organizational cultures in order to obtain the benefits that these can provide for academic and scientific work. The invitation to further contribute to studies that lead to the technological and academic growth of the potential of IRs remains open.

References

Aguillo, I. F., Ortega, J. L., Fernández, M., & Utrilla, A. M. (2010). Indicators for a Webometric ranking of open access repositories. Digital CSIC Sitio web: http://digital.csic.es/bitstream/10261/32190/1/Ranking%20of%20Repositories.pdf

Antonczak, L., Keegan, H., & Cochrane, T. (2016). mLearning and creative practices: A public challenge? *International Journal of Mobile and Blended Learning (IJMBL)*, 8(4), 34–43.

Arlitsch, K., & O'Brien, P. S. (2012). Invisible institutional repositories. *Library Hi Tech*, 30(1), 60–81.

Berquist, T. H. (2015). Open-access institutional repositories: An evolving process? *American Journal of Roentgenology*, 205(3), 467–468.

Berquist, T. H. (2016). Authors and institutional repositories: Working together to reduce the time to final decision. *American Journal of Roentgenology*, 207(1), 1–1.

Björk, B.-C., Laakso, M., Welling, P., & Paetau, P. (2013). Anatomy of green open access. *Journal of the Association for Information Science and Technology*, 65(2), 237–250.

Burns, C. S., Lana, A., & Budd, J. M. (2013). Institutional repositories: Exploration of costs and value. *D-Lib Magazine*, 19(1/2). http://www.dlib.org/dlib/january13/burns/01burns.html.

Cragin, M. H., Palmer, C. L., Carlson, J. R., & Witt, M. (2010). Data sharing, small science and institutional repositories. *Philosophical Transactions of the Royal Society A: Mathematical, Physical and Engineering Sciences*, 368(1926), 4023–4038.

Cullen, R., & Chawner, B. (2011). Institutional repositories, open access, and scholarly communication: A study of conflicting paradigms. *The Journal of Academic Librarianship*, 37(6), 460–470.

Davis, H. C., Carr, L., Hey, J. M. N., Howard, Y., Millard, D., Morris, D., & White, S. (2010). Bootstrapping a culture of sharing to facilitate open educational resources. *IEEE Transactions on Learning Technologies*, 3(2), 96–109.

Dawson, P. H., & Yang, S. Q. (2016). Institutional repositories, open access and copyright: What are the practices and implications? *Science & Technology Libraries*, 35(4), 279–294.

Ezema, I. J. (2011). Building open access institutional repositories for global visibility of Nigerian scholarly publication. *Library Review*, 60(6), 473–485.

Fontes Ferreira, A., & Souza de Silva, A. L. (2015). Institucional repositories: Accesibility to academic research—Public property. In *Proceedings of 2015 International Symposium on Computers in Education (SIIE), IEEE Explore*, 172–175.

Gargouri, Y., Hajjem, C., Lariviére, V., Gingras, Y., Carr, L., Brody, T., & Harnad, S. (2010). Self-selected or mandated, open access increases citation impact for higher quality research. *PLoS ONE, 5*(10), 1–12.

Glassman, M., & Kang, M. J. (2016). Teaching and learning through open source educative processes. *Teaching and Teacher Education, 60*(2016), 281–290.

Gonzalez, L. (2016). Representing serials metadata in institutional repositories. *The Serials Librarian, 70*(1–4), 247–259.

Jean, B. S., Rieh, S. Y., Yakel, E., Markey, K., & Samet, R. (2009). Institutional repositories: What's the use? *Proceedings of the American Society for Information Science and Technology Annual Meeting, 46*(1), 1–5.

Johnson, R., Fosci, M., & Pinfield, S. (2015). Business process costs of implementing "gold" and "green" open access in institutional and national contexts. *Journal of the Association for Information Science and Technology, 67*(9), 2283–2295.

Kim, J. (2011). Motivations of faculty self-archiving in institutional repositories. *The Journal of Academic Librarianship, 37*(3), 246–254.

Koler-Povh, T., Mikoš, M., & Turk, G. (2014). Institutional repository as an important part of scholarly communication. *Library Hi Tech, 32*(3), 423–434.

Laakso, M., & Björk, B. C. (2012). Anatomy of open access publishing: A study of longitudinal development and internal structure. *BMC Medicine, 10*(1), 124.

MacIntyre, R., & Jones, H. (2016). IRUS-UK: Improving understanding of the value and impact of institutional repositories. *The Serials Librarian, 70*(1–4), 100–110.

Marcial, L. H., & Hemminger, B. M. (2010). Scientific data repositories on the web: An initial survey. *Journal of the American Society for Information Science and Technology, 61*(10), 2029–2048.

Pinfield, S., Salter, J., Bath, P. A., Hubbard, B., Millington, P., Anders, J. H. S., & Hussain, A. (2014). Open-access repositories worldwide, 2005–2012: Past growth, current characteristics, and future possibilities. *Journal of the Association for Information Science and Technology, 65*(12), 2404–2421.

Ramírez, M. S. (2015). Acceso abierto y su repercusión en la Sociedad del Conocimiento: Reflexiones de casos prácticos en Latinoamérica. *Education in the Knowledge Society (EKS), 16*(1), 103–118. Available in: http://catedra.ruv.itesm.mx/handle/987654321/873

Rodés-Paragarino, V., Gewerc-Barujel, A., & Llamas-Nistal, M. (2016). Use of repositories of digital educational resources: State-of-the-art review. *IEEE Revista Iberoamericana de Tecnologias del Aprendizaje, 1*(2), 73–78.

Schöpfel, J. (2013). Adding value to electronic theses and dissertations in institutional repositories. *D-Lib Magazine, 19*(3–4).

Shukair, G., Loutas, N., Peristeras, V., & Sklarß, S. (2013). Towards semantically interoperable metadata repositories: The asset description metadata schema. *Computers in Industry, 64*(1), 10–18.

Sotiriou, S., Riviou, K., Cherouvis, S., Chelioti, E., & Bogner, F. X. (2016). Introducing large-scale innovation in schools. *Journal of Science Education and Technology, 25*(4), 541–549.

Tzoc, E. (2016). Institutional repository software platforms at undergraduate libraries in the United States. *College & Undergraduate Libraries, 23*(2), 184–192.

Wallace, J. M. (2011). PEER: Green open access—Insight and evidence. *Learned Publishing, 24*(4), 267–277.

Way, D. (2010). The open access availability of library and information science literature. *College and Research Libraries, 71*(4), 302–309.

Wu, M. (2015). The future of institutional repositories at small academic institutions: Analysis and insights. *D-Lib Magazine, 21.*

Xia, J., Gilchrist, S. B., Smith, N. X., Kingery, J. A., Radecki, J. R., Wilhelm, M. L., Harrison, K. C. et al. (2012). A review of open access self-archiving mandate policies. *portal: Libraries and the Academy, 12*(1), 85–102.

APPLICATION OF SCIENTOMETRICS TO UNIVERSITY COMPETITIVENESS AND WORLD-CLASS UNIVERSITIES

II

Chapter 6

Academic Ranking of World Universities (ARWU): Methodologies and Trends

Yan Wu and Nian Cai Liu

Graduate School of Education, Shanghai Jiao Tong University, Min Hang, Shanghai, China

Contents

6.1 Introduction

6.1.1 ARWU: Born for the Chinese Dream

Achieving the Chinese nation's bright prospect on the road to revival is the dream for many generations. World-class universities are essential in developing a nation's competitiveness in the global knowledge economy. So, the Chinese government initiated the strategic policy of building world-class universities, which resulted in the 985 Project. In the first phase, nine top Chinese universities were selected to build world-class universities by the 985 Project, which are Peking University, Tsinghua University, Shanghai Jiao Tong University (SJTU), Fudan University, Zhejiang University, Nanjing University, University of Science and Technology of China, Xi'an Jiao Tong University, and Harbin Institute of Technology. Right after its selection, SJTU assigned our department, the Office of Strategic Planning of the university, to be responsible for making the strategic plan.

During the process, there were many questions asked by ourselves. For example, what is the definition of a world-class university? How many world-class universities should there be all over the world? What are the positions of top Chinese universities in the world higher education system? How can top Chinese universities reduce their gap with world-class universities? In order to answer these questions, we started to benchmark top Chinese universities with world-class universities. To answer these questions, we not only did a qualitative study but also a quantitative comparison. From 1999 to 2001, we worked on the project of benchmarking top Chinese universities with four groups of U.S. universities, from the very top to the less-known research universities, according to a wide spectrum of indicators of academic or research performance. The results of these comparisons and analyses were used in the planning process, including that top Chinese universities were estimated to be approximately in the position of 200 to 300 in the world. Eventually, a consultation report was provided to the Ministry of Education of China and reviewed favorably by the government officials.

The internationalization of higher education worldwide was growing rapidly in the early twenty-first century. Universities, governments, and other stakeholders not only in China but also in the rest of the world were interested in the quantitative comparison of world universities. So, we decided to make a real ranking of global universities. After another 2 years of benchmarking work, the first ARWU ranking was released in 2003 on the website http://www.shanghairanking.com/ and became the first multi-indicator ranking of global universities. Since then, it has been updated and published annually. By now, ARWU has been providing trustworthy performance information on universities in different countries for 14 years.

6.1.2 Drawing Global Attention

Ever since its publication, ARWU has attracted a great deal of attention from universities, governments, and public media worldwide. The EU Research Headlines reported ARWU that "the universities were carefully evaluated using several indicators of research performance" (European Commission, 2003). As Figure 6.1 indicates, about one-fourth of the major media in the United States, the United Kingdom, Germany, France, Australia, Japan, South Korea, and other countries reported ARWU rankings in 2014.

Universities may take the ARWU rankings as effective tools in building and maintaining reputation, which are important to attract talents and resources and to gain the support of the general public. The University of California, University of Cambridge, University of Tokyo, the Australian National University, and a number of world famous universities in their own annual reports or home pages also report or quote ARWU. From 2003 to 2014, hundreds of universities cited them in their reports (Figure 6.2).

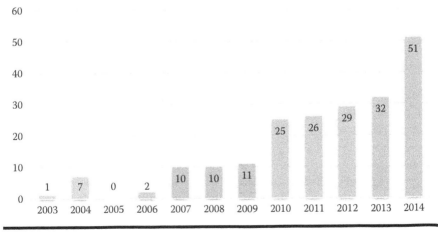

Figure 6.1 The number of major media that reported ARWU rankings worldwide.

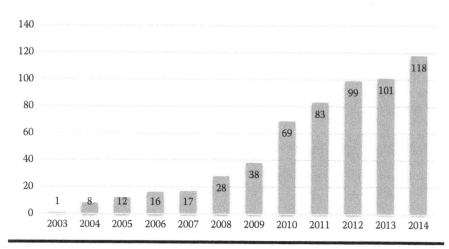

Figure 6.2 The number of top 200 universities that reported or quoted ARWU (2003–2014).

6.2 Methodologies and Impacts

6.2.1 Methodologies

The design of the indicator system reflects the rankers' conception about the universities, as well as the original purpose about the usage of rankings. As described previously, ARWU originated from identifying the world-class universities and finding the gap between Chinese universities and world-class universities, particularly in terms of academic or research performance. In addition, the availability of internationally comparable data was considered. Therefore, the indicators for measuring academic excellence were chosen, including the winners of Nobel Prizes that did cutting-edge research; Highly Cited Researchers, who represented some of the world's most influential scientific minds; and articles published in the journals *Nature* and *Science*, which were supposed to be groundbreaking research.

ARWU uses six objective indicators to rank world universities. The detailed indicators, weights, and data sources are provided in Table 6.1.

ARWU considers every university that has any Nobel laureates, Fields Medals, Highly Cited Researchers, or articles published in *Nature* or *Science*. In addition, major universities of every country with a significant amount of articles indexed by SCIE and SSCI are also included. In total, about 1,200 universities have actually been ranked, and the best 500 universities are published on the ARWU website.

Table 6.1 Indicators, Weights, and Data Sources for ARWU

Criterion	Indicator	Weight (%)	Data Sources
Quality of education	*Alumni:* Alumni of an institution winning Nobel Prizes and Fields Medals	10	http://www.nobelprize.org/ http://www.mathunion.org /index.php?id=prizewinners
Quality of faculty	*Award:* Staff of an institution winning Nobel Prizes and Fields Medals	20	
	HiCi: Highly Cited Researchers in 21 broad subject categories	20	http://www.highlycited.com/
Research output	*N&S:* Papers published in *Nature* and *Science*[a]	20	http://www .webofknowledge.com/
	PUB: Papers indexed in *Science Citation Index Expanded* (SCIE) and *Social Science Citation Index* (SSCI)	20	http://www .webofknowledge.com/
Per capita performance	*PCP:* Per capita academic performance of an institution	10	National agencies such as the National Ministry of Education and National Bureau of Statistics
Total		100	

Source: ShanghaiRanking Consultancy, Shanghai, http://www.shanghairanking.com /ARWU-Methodology-2016.html.

[a] For universities specializing in humanities and social sciences such as London School of Economics, N&S is not considered, and the weight of N&S is relocated to other indicators.

6.2.2 Features

ARWU is well known for its special features since its first publication. Simon Marginson (2014) wrote that "in terms of data quality, ARWU has no areas of fundamental weakness. Its strong points are materiality, objectivity and reliance on external data. These factors have generated a high level of trust in the ARWU (it also has first mover advantage as the first credible research ranking)."

First, ARWU uses objective indicators only, which distinguishes it from other major global rankings. The objective indicators used in ARWU are more comparable and reliable than survey data, which can be affected by some psychological effects such as the *halo effect* (Woodhouse, 2008) and *leniency effect* (Van Dyke, 2008).

Second, ARWU uses third-party objective data only. There are no data collected directly from the individual universities ranked, and cleanup of data is performed whenever necessary. As a result, there is no possibility for universities to manipulate data as in the case when institutions are requested to submit to ranking organizations.

Third, the methodology of ARWU is transparent and open to suggestions and criticisms. Every university can verify their positions by comparing indicator scores with those of their peer institutions. In fact, the majority of institutions in our list have verified with us. A large number of institutions not in the top 500 list have also checked with us about their performance.

Furthermore, we kept the changes in ranking methodology to a minimum. As a matter of fact, there have been no significant changes of ARWU methodology since 2004. Therefore, various stakeholders including universities and governments can compare the performances of particular universities or countries across the years.

6.2.3 Impact

As the first multi-indicator global university ranking, ARWU started the international ranking phenomenon. Nearly 1.5 years after the first publication of ARWU, the Times Higher Education (THE) Supplement published its "World University Rankings" in 2004 with the comment that "domestic league tables are controversial enough, but there are extra pitfalls associated with international comparisons" (Times Higher Education, 2004). Until now, more than a dozen global rankings of universities have been published. Ellen Hazelkorn (2014) wrote, "today, global rankings have become the simple (and simplistic) tool of choice for a wide range of stakeholders on the presumption (as yet unproven) that they provide a good measure of quality … ARWU marked the era of global rankings, despite being developed to highlight the position of Chinese universities vis-a-vis competitor universities and being entirely focused on research, it has effectively become the 'gold standard.'"

The director of the Center for the Study of State and Society, Ana García de Fanelli (2012), wrote, "My personal view is that global rankings play an important role in guiding the development of research universities and in the design of public policy. ... This was, in fact, the public policy goal of the ranking initially launched by Jiao Tong University in China." After its publication, ARWU has triggered reflections on the current situation and future development of higher education by various governments and universities. For example, Martin Enserink (2007) wrote that "France's poor showing in the Shanghai rankings—it had only two universities in the first top 100—helped trigger a national debate about higher education that resulted in a new law, passed last month, giving universities more freedom." In 2012, Russian president Vladimir Putin, through a presidential decree, promulgated and implemented a project called the 5-100 plan, which aims to enhance the international competitiveness of Russian universities and bring at least five Russian universities by 2020 ranking among the top 100 global universities (Russian Academic Excellence Project, 2012). In 2013, Japanese prime minister Shinzo Abe and his cabinet announced Japan's revitalization strategy and policy framework that by 2023, at least 10 universities would be promoted into the global top 100 (Smith, 2013). In addition, France, South Korea, Taiwan, Singapore, Malaysia, and other Asian countries or regions are using university rankings to adjust the higher education strategy to enhance their international competitiveness.

The value of ARWU has been widely recognized to be a credible and transparent tool for monitoring universities' global competitiveness. Universities frequently make reference to ARWU for various purposes, including setting goals in their strategic plans, positioning themselves in the world, identifying strengths and weaknesses for strategic planning, selecting target universities for benchmarking and international collaboration, improving visibility and building reputation, recruiting academic staff and students, awarding scholarship for international studies, and so forth. For example, the University of Manchester (2011) in the United Kingdom set its goal into ARWU top 25 by 2020 in its latest strategic plan. The University of Melbourne (2014) is committed to entering ARWU top 40 within the next decade (2013–2023), and the University of Western Australia (2010) is committed to reaching ARWU top 50 by 2050. As the example to award scholarship, Zayed University (2014) has the rule that "masters' students have a choice of several universities and disciplines in UAE and at foreign universities included in the top 200 Academic Ranking of World Universities."

6.3 Changes and Trends of National and Regional Performance in ARWU

The methodology of ARWU is transparent and stable, so that one can compare the performance of particular universities or countries throughout the years. In this section, the changes of performance of universities and countries over the past 13 years are provided.

6.3.1 Performance of Countries and Regions in ARWU 2016

According to Table 6.2, higher education in the United States still takes the lead, with 137 universities belonging to the top 500 group, accounting for 27.4% of all top 500 universities, and 15 out of these 137 universities are in the most leading positions—the top 20. In terms of the amount of the top 500 universities, the United Kingdom ranks fourth, and with three universities in the top 20 list, the United Kingdom is regarded as the second education powerhouse. Additionally, one Japanese university and one Swiss university are also in the top 20 group.

Brazil performs the best among Latin American countries, with six universities in the top 500. South Africa is the best among African countries, with four universities in the top 500. Among Arab countries, Saudi Arabia takes the leading position, with four universities in the top 500. New Zealand as another country in Oceania has good performance with four universities in the top 500. Although there are only two Singaporean universities on the list, they rank in the top 100 list and the top 200, respectively.

There were six universities making their first appearance in the top 100 list in 2016, including Tsinghua University, Peking University, Monash University, the National University of Singapore, Mayo Medical School, and the University of Texas MD Anderson Cancer Center. There were 16 new institutions in the top 500 list in ARWU 2016, with China, Australia, Italy, and Portugal each having two new universities.

Table 6.3 shows the statistics of the number of universities listed in ARWU's top 100, 101–200, 201–300, 301–400, and 401–500 by region and subregion. There are more universities listed in the top 100 in the United States than those in Europe, but the total number of the top 500 ones in Europe is the largest in the world. The performance of universities in four subregions of Europe varies significantly, while the performance of universities in four subregions of Asia has much bigger differences, with the majority of universities from East Asia.

6.3.2 Performance Trends of Selected Countries and Regions from 2004 to 2016

Figure 6.3 presents the trends of the number of universities listed in ARWU's top 500 in selected countries and regions. The top 500 universities in the United States and continental Europe far outnumber other countries and regions, constituting 66% and 60% of the total top 500 universities in 2004 and 2016, respectively.

The number of the top 500 universities from both the United States and Japan shows a continuing decline, with an annual average decrease of 2.5 U.S. universities

Table 6.2 Number of Universities Listed ARWU by Country (2016)

Country	Top 20	Top 21–100	Top 101–200	Top 201–300	Top 301–400	Top 401–500	Total
U.S.	15	35	21	27	21	18	137
China	–	2	10	12	16	14	54
Germany	–	3	11	7	6	11	38
UK	3	5	13	7	5	4	37
Australia	–	6	2	6	7	2	23
France	–	3	6	4	5	4	22
Canada	–	4	2	7	3	3	19
Italy	–	–	2	6	3	8	19
Japan	1	3	2	3	3	4	16
Netherlands	–	3	6	2	1	–	12
Spain	–	–	1	2	6	3	12
Sweden	–	3	2	3	2	1	11
South Korea	–	–	3	3	3	2	11
Switzerland	1	3	2	1	1	–	8
Belgium	–	2	2	2	1	–	7
Brazil	–	–	1	–	3	2	6
Denmark	–	2	1	1	1	–	5
Israel	–	2	2	–	–	1	5
Finland	–	1	–	–	1	3	5
Austria	–	–	2	1	–	2	5
Portugal	–	–	1	–	1	3	5

Source: ShanghaiRanking Consultancy, Shanghai, http://www.shanghairanking .com/ARWU-Statistics-2016.html.

Note: Only the countries with more than five top 500 universities are listed in the table.

Table 6.3 Number of Universities Listed in ARWU by Region and Subregion (2016)

Region and Subregion	Top 100	Top 101–200	Top 201–300	Top 301–400	Top 401–500	Total
North America	54	24	34	24	21	157
Europe	31	51	39	37	46	204
Western Europe	15	29	17	14	17	92
Northern Europe	15	18	12	11	9	65
Eastern Europe	1	0	1	1	3	6
Southern Europe	0	4	9	11	17	41
Asia	9	20	19	25	26	99
East Asia	6	15	18	22	20	81
West Asia	2	4	1	1	2	10
Southeast Asia	1	1	0	0	3	5
South Asia	0	0	0	2	1	3
Oceania	6	3	6	10	2	27
South America	0	2	0	4	2	8
Africa	0	0	2	0	3	5

Source: ShanghaiRanking Consultancy, Shanghai, http://www.shanghairanking .com.

and 1.5 Japanese universities. By contrast, China has witnessed an average increase of two universities in the top 500 list per year, and Oceania and the Middle East each have an average of one new university entering the top 500 list every year. Basically, the number of the top 500 universities in continental Europe and Latin America remains steady, with two more universities in the top 500 list, respectively, from 2004 to 2016.

Figure 6.4 illustrates the trends of the number of universities listed in ARWU's top 100 in selected countries and regions. When compared with Figure 6.3, Latin American countries disappeared from the figure because they have no university listed in the top 100. The top 100 universities from the United States and continental Europe significantly outnumber those in other countries and regions, but

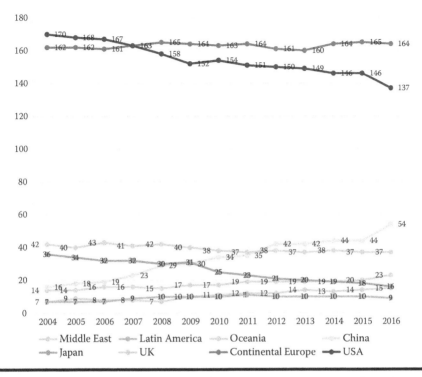

Figure 6.3 Trends of universities listed in ARWU's top 500 by selected country and region (2004–2016). Continental Europe is made up of every country in Europe except the Channel Islands, Iceland, Ireland, the United Kingdom, and the Isle of Man. In Europe, continental Europe is often referred to as mainland Europe (Reference.com, 2016). The Middle East is made up of every country in Western Asia and Egypt (Wikipedia, 2016). (Data from ShanghaiRanking Consultancy, ARWU, 2004–2016, Shanghai.)

they show overall declining trends with fluctuations over the years. The decrease of the top 100 universities from continental Europe is mainly a result of the reduced number of the top 100 universities from Germany and France. The number of the top 100 universities from the United Kingdom and Japan is significant, but shows overall declining trends.

The performance of Oceania in the top 100 list is similar to its performance in the top 500 list. It is worthy to mention that all Oceanian universities in the top 100 list are Australian universities. Also, all top 100 universities in the Middle East belong to Israel.

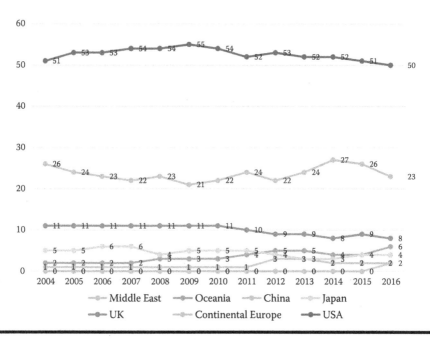

Figure 6.4 Trends of universities listed in ARWU's top 100 by selected country and region (2004–2016). (Data from ShanghaiRanking Consultancy, ARWU, 2004–2016, Shanghai.)

6.3.3 Performance Changes of Countries from 2004 to 2016

Tables 6.4 and 6.5 demonstrate, respectively, the increases and decreases in the number of universities in the top 500 list from 2004 to 2016 by country. The total number of countries having universities in the top 500 list increased from 35 in 2004 to 43 in 2016. Nine countries, including Slovenia, Saudi Arabia, Iran, Malaysia, Croatia, Egypt, Serbia, Estonia, and Turkey, not in the top 500 list of 2004 appeared in the top 500 list of 2016, while Hungary is the only country which was in the top 500 list in 2004 but disappeared from the list in 2016.

Among 18 countries with an increased number of universities in the top 500 list, China takes the leading place, and Australia is second after China. Moreover, four more Australian universities appeared in the top 100 list of 2016 as compared with 2004. Universities in Portugal and Spain also perform well with an additional eight universities in the top 500.

There were 10 countries with a decreased number of the top 500 universities. These are mainly developed countries, including the United States, the United Kingdom, and Japan. Thirty-three U.S. universities, 20 Japanese universities, 5 British universities, and 5 German universities are out of the list.

Table 6.4   Increases in the Number of Universities Listed in ARWU from 2004 to 2016 by Country

Country	Top 20	Top 21–100	Top 101–200	Top 201–300	Top 301–400	Top 401–500	Total
China	–	2	9	7	9	11	38
Australia	–	4	–2	4	4	–1	9
Portugal	–	–	1	–	1	3	5
Saudi Arabia	–	–	2	1	1	–	4
Malaysia	–	–	–	–	–	3	3
Spain	–	–	–	1	4	–2	3
South Korea	–	–	2	2	1	–2	3
Brazil	–	–	–	–	1	1	2
Iran	–	–	–	–	1	1	2
Sweden	–	–1	1	–1	2	–	1
Russia	–	–	–	–	–	1	1
Croatia	–	–	–	–	–	1	1
Egypt	–	–	–	–	–	1	1
Estonia	–	–	–	–	–	1	1
Slovenia	–	–	–	–	–	1	1
Turkey	–	–	–	–	–	1	1
Serbia	–	–	–	1	–	–	1
New Zealand	–	–	1	–2	3	–1	1

Source: ShanghaiRanking Consultancy, Shanghai, http://www.shanghairanking.com.

Among countries with the total number of the top 500 universities unchanged from 2004 to 2016, some made significant progress in the actual positions. For example, one Swiss university entered the top 20 list and two Belgian universities jumped from the top 101–200 to the top 100. Denmark, Singapore, and the Netherlands each had one university climbing from the top 101–200 to the top 100. On the other hand, an Austrian university and a French university dropped out of the top 100 list.

Table 6.6 illustrates the year-by-year changes in the total number of universities listed in ARWU's top 100 from 2004 to 2016 by country. There are 18 countries

Table 6.5   Decreases and No Change in the Number of Universities Listed in ARWU from 2004 to 2016 by Country

Country	Top 20	Top 21–100	Top 101–200	Top 201–300	Top 301–400	Top 401–500	Total
U.S.	–2	1	–18	–2	2	–14	–33
Japan	–	–1	–2	–1	–10	–6	–20
UK	1	–4	6	–4	–1	–3	–5
Germany	–	–3	2	–2	–5	5	–5
Canada	–	–	–3	–	–	–1	–4
Italy	–	–1	–2	1	–3	2	–3
Hungary	–	–	–	–1	–	–2	–3
Israel	–	1	–	–1	–2	–	–2
India	–	–	–	–1	1	–2	–2
Norway	–	–	1	1	–2	–1	–1
Austria	–	–1	2	–1	–1	1	–
France	–	–1	3	–2	–2	2	–
Finland	–	–	–	–1	–1	2	–
Poland	–	–	–	–	–2	2	–
Greece	–	–	–	–	–1	1	–
Chile	–	–	–	–	–	–	–
Mexico	–	–	–	–	–	–	–
South Africa	–	–	–	1	–1	–	–
Czech	–	–	–	1	–1	–	–
Argentina	–	–	1	–1	–	–	–
Ireland	–	–	1	–1	2	–2	–
Denmark	–	1	–1	–	–	–	–
Singapore	–	1	–	–	–	–1	–
Netherlands	–	1	1	–1	–	–1	–
Belgium	–	2	–2	–	–	–	–
Switzerland	1	–	–1	1	–	–1	–

Source: ShanghaiRanking Consultancy, Shanghai, http://www.shanghairanking.com.

Table 6.6 Year-by-Year Changes of Universities in ARWU's Top 100 from 2004 to 2016 by Country

Country	2016	2015	2014	2013	2012	2011	2010	2009	2008	2007	2006	2005	2004
U.S.	50	51	52	52	53	52	54	55	54	54	53	53	51
UK	8	9	8	9	9	10	11	11	11	11	11	11	11
Australia	6	4	4	5	5	4	3	3	3	2	2	2	2
Japan	4	4	3	3	4	5	5	5	4	6	6	5	5
Canada	4	4	4	4	4	4	4	4	4	4	4	4	4
Switzerland	4	4	5	4	4	4	3	3	3	3	3	3	3
Germany	3	4	4	4	4	6	5	5	6	6	5	5	7
France	3	4	4	4	3	3	3	3	3	3	4	4	4
Sweden	3	3	3	3	3	3	3	3	4	4	4	4	4
Netherlands	3	4	4	3	2	2	2	2	2	2	2	2	2
Denmark	2	2	2	2	2	2	2	2	2	1	1	1	1

(Continued)

Table 6.6 (Continued) Year-by-Year Changes of Universities in ARWU's Top 100 from 2004 to 2016 by Country

Country	2016	2015	2014	2013	2012	2011	2010	2009	2008	2007	2006	2005	2004
Israel	2	2	2	3	3	1	1	1	1	1	1	1	1
Belgium	2	2	2	1	1	1	1						
China	2	–	–	–	–	–	–	–	–	–	–	–	–
Finland	1	1	1	1	1	1	1	1	1	1	1	1	1
Norway	1	1	1	1	1	1	1	1	1	1	1	1	1
Russia	1	1	1	1	1	1	1	1	1	1	1	1	1
Singapore	1	–	–	–	–	–	–	–	–	–	–	–	–
Austria	–	–	–	–	–	–	–	–	–	–	–	1	1
Italy	–	–	–	–	–	–	–	–	–	–	1	1	1
Total	100	100	100	100	100	100	100	100	100	100	100	100	100

Source: ShanghaiRanking Consultancy, Shanghai, http://www.shanghairanking.com.

in 2016 as compared with 17 countries in 2004. Belgium made its first appearance in the top 100 list in 2010. For the first time in 2016, China and Singapore gained positions in the top 100. Austria and Italy lost their top 100 universities in 2006 and 2007, respectively.

Among countries which have an increased number of the top 100 universities, Australia has made impressive progress, with the number of the top 100 universities increased from two in 2004 to six in 2016. Switzerland, Netherlands, Denmark, Belgium, Israel, China, and Singapore also performed well.

By and large, the top 100 universities from the United Kingdom and Germany continued to decrease since 2004. The number of the United Kingdom's top 100 universities has decreased from 11 in 2004 to nine in 2016, while that of Germany's top 100 universities has dropped from seven to three. Countries with a decreased number of the top 100 universities also include France and Sweden.

6.4 Changes and Trends of Institutional Performance in ARWU

In this section, the changes and trends of institutional performance in ARWU from 2004 to 2016 will be presented and discussed.

6.4.1 Statistics on the Numbers of Institutions in ARWU from 2004 to 2016

Table 6.7 presents the statistics on the numbers of institutions in ARWU top 500, top 100, and top 20 lists between 2004 and 2016.

Between 2004 and 2016, 643 universities have appeared at least once in the top 500 list; they are from the United States (183, 28.5%), China (57, 8.9%), the

Table 6.7   Numbers of Institutions in ARWU between 2004 and 2016

	Number of Institutions
Institutions appearing in the top 500 list at least once	643
Institutions staying in the top 500 list for all the years	372
Institutions appearing in the top 100 list at least once	129
Institutions staying in the top 100 list for all the years	81
Institutions appearing in the top 20 list at least once	23
Institutions staying in the top 20 list for all the years	16

United Kingdom (48, 7.5%), Germany (47, 7.4%), Japan (38, 5.9%), France (31, 4.8%), Italy (28, 4.4%), Australia (25, 3.9%), Canada (24, 3.7%), and Spain (15, 2.3%). There are 372 universities out of the total 643 universities that have kept their positions as the top 500 universities for all the years ranked. They are from the United States (34.1%), Germany (9.1%), the United Kingdom (9.1%), Canada (5.1%), China (4.3%), France (3.8%), Italy (3.5%), Australia (3.2%), Japan (3.2%), and the Netherlands (3.2%).

Between 2004 and 2016, 129 universities from 20 countries have appeared at least once in the top 100 list. There are 61 U.S. universities that account for nearly 50% of the total, and the second to sixth largest number of universities are 12 British universities (9.3%), eight German universities (6.2%), six Australian universities (4.7%), six Japanese universities (4.7%), and five Swiss universities (3.9%). Out of a total of 129 universities, 81 are staying in the top 100 list for all the years ranked, including 46 U.S. universities (56.8%), eight British universities (9.9%), four Canadian universities (4.9%), three French universities (3.7%), three German universities (3.7%), three Japanese universities (3.7%), and three Swedish universities (3.7%).

Between 2004 and 2016, 23 universities have appeared at least once in the top 20 list, including 18 U.S. universities (78.3%), three British universities (13.0%), one Japanese university, and one Swiss university. Sixteen universities out of the total 23 are staying in the top 20 list for all the years ranked, with 14 from the United States and two from the United Kingdom.

6.4.2 Institutions with Great Improvements in ARWU from 2004 to 2016

Table 6.8 shows the 20 universities which improved the most among the 643 universities which have appeared at least once in the top 500 list from 2004 to 2016.

Among the 20 universities of 643 universities which have appeared at the least once in the top 500 list, there is only one European university from Portugal and the rest are universities from Asia and Oceania, including 11 Chinese universities, three Australian universities, three Saudi Arabian universities, one Singaporean university, and one Israeli university. Universities in America and Africa have not shown up.

Among these universities, four have been entered in the top 100, namely, Tsinghua University (China), Technion-Israel Institute of Technology (Israel), Peking University (China), and Monash University (Australia). Ten have entered in the top 101–200 list.

Table 6.9 shows the 10 universities with the biggest rise in the ranking positions among the previously mentioned 129 universities, which have appeared at least once in the top 100 list between 2004 and 2016, including three Australian universities, two Chinese universities, two Swiss universities, one Israeli university, one Belgian university, and one British university.

Table 6.8 Twenty Universities That Improved the Most in ARWU's Top 500 List (2004 vs. 2016)

Institution	Country	Ranks in 2016	Ranks in 2004	First Year in Top 500
King Abdulaziz University	Saudi Arabia	101–150	500+	2012
King Saud University	Saudi Arabia	101–150	500+	2010
Peking University	China	71	201–300	2003
King Abdullah University of Science and Technology	Saudi Arabia	201–300	500+	2013
Southeast University	China	201–300	500+	2010
Sun Yat-sen University	China	151–200	500+	2007
Tsinghua University	China	59	201–300	2003
Xian Jiao Tong University	China	151–200	500+	2010
Shanghai Jiao Tong University	China	101–150	401–500	2003
China Medical University (Taiwan)	China	151–200	500+	2012
Harbin Institute of Technology	China	151–200	500+	2008
Nanyang Technological University	Singapore	101–150	401–500	2003
Technion–Israel Institute of Technology	Israel	69	201–300	2003
University of Lisbon	Portugal	151–200	500+	2007
Sichuan University	China	201–300	500+	2008
Deakin University	Australia	201–300	500+	2014
Curtin University	Australia	201–300	500+	2009
Zhejiang University	China	101–150	301–400	2003
Huazhong University of Science and Technology	China	201–300	500+	2008
Monash University	Australia	79	201–300	2003

Source: ShanghaiRanking Consultancy, ARWU, 2004 and 2016, Shanghai.

Table 6.9   Ten Universities That Improved the Most in ARWU's Top 100 (2004 vs. 2016)

	Country	Ranks in 2016	Ranks in 2004	First Year in Top 100
Peking University	China	71	201–300	2016
Tsinghua University	China	59	201–300	2016
Technion–Israel Institute of Technology	Israel	69	201–300	2012
Monash University	Australia	79	201–300	2016
The University of Queensland	Australia	55	101–150	2011
University of Geneva	Switzerland	53	101–150	2011
The University of Manchester	UK	35	78	2003
Swiss Federal Institute of Technology Lausanne	Switzerland	92	151–200	2014
The University of Melbourne	Australia	40	83	2003
Ghent University	Belgium	62	101–150	2010

Source: ShanghaiRanking Consultancy, ARWU, 2004 and 2016, Shanghai.

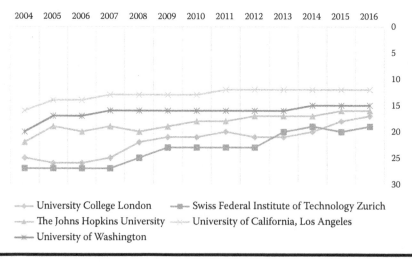

—◆— University College London   —■— Swiss Federal Institute of Technology Zurich
—▲— The Johns Hopkins University   —✕— University of California, Los Angeles
—✱— University of Washington

Figure 6.5   Five universities that have maintained a rising trend in the ARWU's top 20 list.

Figure 6.5 reveals five out of the previously mentioned 23 universities that have maintained a rising trend in the top 20 ranking. With the exception of the University of California–Los Angeles (United States), which stayed in the top 20 for all the years ranked, the other four universities climbed to the top 20 list from lower positions, with University College London (United Kingdom) and Swiss Federal Institute of Technology Zurich (Switzerland) showing the fastest rise.

6.4.3 Institutions with Major Declines in ARWU from 2004 to 2016

Table 6.10 indicates the 20 universities whose ranking positions dropped the most among all universities that have ever appeared in the top 500 list.

These universities consist of 10 U.S. universities, eight Japanese universities, one German university, and one British university. Universities whose ranking positions dropped the most were Yamaguchi University (Japan), Tokyo Institute of Technology (Japan), and Polytechnic Institute of New York University (United States). In 2016, there were seven out of these 20 universities still in the top 500, and the rest were out of the top 500.

Among the 372 universities that have kept their positions as the top 500 universities from 2004 to 2016, 10 may move out from the top 500 list in the near future, including three U.S. universities, two Italian universities, one South African university, one French university, one German university, one Israeli university, and one Finnish university. As can be seen in Figure 6.6, the University of Texas Medical Branch (UTMB Health, United States) has gone through the most obvious decline in the ranking positions.

Table 6.11 describes universities with the biggest decline of the ranking positions among the previously mentioned 129 universities, which have been staying in the top 100 list. The 10 fastest dropping universities include five U.S. universities, two Japanese universities, one Austrian university, one Italian university, and one British university.

Five out of these 81 universities which have stayed in the top 100 list for all the years ranked may be out of the top 100 group in the near future (Figure 6.7), including two U.S. universities, one Japanese university, one Russian university, and one Netherlands university.

Figure 6.8 presents universities that have maintained a downward trend in the top 20 list. The University of California–San Francisco, University of Michigan–Ann Arbor, and University of Wisconsin–Madison have dropped out of the top 20 list.

Table 6.10   Twenty Universities That Declined the Most in ARWU's Top 500 List (2004 vs. 2016)

Institution	Country	Ranks in 2016	Ranks in 2004	First Year Not in Top 500
Yamaguchi University	Japan	500+	401–500	2012
Tokyo Institute of Technology	Japan	201–300	101–150	–
Polytechnic Institute of New York University	U.S.	500+	401–500	2005
Medical College of Georgia	U.S.	500+	401–500	2009
Rensselaer Polytechnic Institute	U.S.	401–500	151–200	–
Himeji Institute of Technology	Japan	500+	401–500	2005
University of Rochester	U.S.	101–150	52	–
Syracuse University	U.S.	500+	201–300	2016
Gunma University	Japan	500+	301–400	2011
Tokyo University of Agriculture and Technology	Japan	500+	301–400	2011
University of Greifswald	Germany	500+	301–400	2012
Gifu University	Japan	500+	301–400	2010
Rutgers, The State University of New Jersey–New Brunswick	U.S.	97	44	–
University of Bradford	UK	500+	401–500	2007
Howard University	U.S.	500+	401–500	2007
Tokyo Metropolitan University	Japan	500+	301–400	2009
Washington State University	U.S.	401–500	151–200	—
University of Tennessee–Knoxville	U.S.	301–400	101–150	—
Tulane University	U.S.	500+	201–300	–
Tohoku University	Japan	101–150	70	–

Source: ShanghaiRanking Consultancy, ARWU, 2004 and 2016, Shanghai.

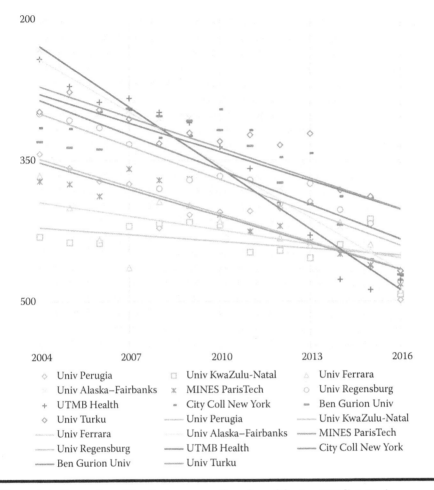

Figure 6.6 Ten universities that are near the edge of the top 500 list. Various symbols represent the actual ranking position of a university, and the oblique lines indicate the changing trend of a university's ranking. (Data from ShanghaiRanking Consultancy, ARWU, 2004–2016, Shanghai.)

6.5 Concluding Remarks

6.5.1 Future of ARWU

As the first multi-indicator ranking of global universities, ARWU has been providing reliable and scientific performance comparison of universities in the world for the past 14 years. As we did in the past 13 years, we will continue to update ARWU annually and keep its methodology stable in the future.

Table 6.11   Ten Universities That Declined the Most in ARWU's Top 100 (2004 vs. 2016)

	Country	Ranks in 2016	Ranks in 2004	First Year Not in Top 100
University of Rochester	U.S.	101–150	52	2015
Rutgers, The State University of New Jersey–New Brunswick	U.S.	97	44	–
Tohoku University	Japan	101–150	70	2012
University of Vienna	Austria	151–200	86	2006
Osaka University	Japan	99	54	–
The University of Sheffield	UK	101–150	69	2012
Pennsylvania State University–University Park	U.S.	77	43	–
University of California, Davis	U.S.	75	42	–
University of Roma–La Sapienza	Italy	151–200	94	2006
Case Western Reserve University	U.S.	101–150	65	2014

Source: ShanghaiRanking Consultancy, ARWU, 2004 and 2016, Shanghai.

6.5.2 New Efforts

We have been exploring the possibility to rank higher education institutions from more aspects, for example, ranking of globally comparable subjects. In 2016, a new subject ranking with a cluster of engineering subjects is published. This ranking utilizes our latest research results, such as more international scientific awards and more internationally renowned scholars.

6.5.3 Cautions

In the era of information explosion, university rankings gain extensive popularity, because rankings can be regarded as a simple and convenient way to obtain information. However, rankings are controversial to some degree. Therefore, people should pay more attention to the methodologies and indicators of rankings so as to understand and apply their results more accurately.

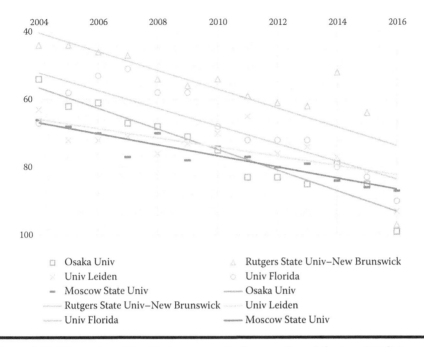

Figure 6.7 Five universities that are near the lower edge of the top 100 list.

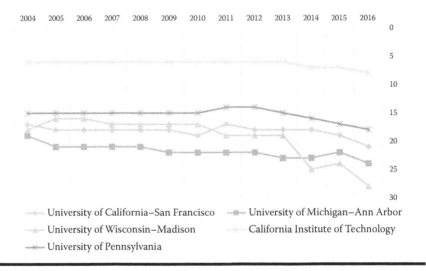

Figure 6.8 Five universities that have a downward trend in ARWU's top 20 list.

Acknowledgments

The authors acknowledge the financial support of the National Social Sciences Foundation of China (Grant No. BIA120074).

References

de Fanelli, A. G. (2012). More thoughts on higher education rankings and Latin American universities. Retrieved from https://www.insidehighered.com/blogs/world-view /more-thoughts-higher-education-rankings-and-latin-american-universities. 2016-11-06

Enserink, M. (2007). Who ranks the university rankers? *Science, 317*(5841), 1026–1028.

European Commission. (2003). Chinese study ranks world's top 500 universities. Retrieved from http://ec.europa.eu/research/headlines/news/article_03_12_31_en.html. 8-25

Hazelkorn, E. (2014). Reflections on a decade of global rankings: What we've learned and outstanding issues. *European Journal of Education, 49*(1), 12–28.

Marginson, S. (2014). University rankings and social science. *European Journal of Education, 49*(1), 45–59.

Reference.com. (2016). What countries make up continental Europe? Retrieved from https://www.reference.com/geography/countries-make-up-continental-europe -19b6e4932b020d69#. 2016-09-22

Russian Academic Excellence Project. (2012). First XV: How the Project 5-100 stars are transforming the Russian academy. Retrieved from http://5top100.ru/upload/iblock /6a0/6a04e0f4f0abb1fee842f7b32a9a6e4a.pdf

Smith, G. (2013). Japan facing mixed future on education policy. *Education News*. Retrieved from http://www.educationnews.org/international-uk/japan-facing-mixed-future-on -education-policy/#sthash.ahCkosYX.dpuf. 2014-10-23. 2016-11-06

Times Higher Education. (2004). Top performers on the global stage take a bow. *The Times Higher Education Supplement*. Retrieved from https://www.timeshighereducation .com/features/editorial-top-performers-on-the-global-stage-take-a-bow/192258 .article

University of Manchester. (2011). Manchester 2020: The strategic plan for the University of Manchester. Retrieved from http://documents.manchester.ac.uk/display.aspx?DocID =11953

University of Melbourne. (2014). Growing esteem 2014. Retrieved from http://growing esteem.unimelb.edu.au/documents/UoM_Growing_Esteem_2014_GreenPaper.pdf

University of Western Australia. (2010). The Campus Plan 2010. Retrieved from http:// www.cm.uwa.edu.au/plan/campus-plan-2010/vc-message

Van Dyke, N. (2008). Self- and peer-assessment disparities in university ranking schemes. *Higher Education in Europe, 33*(2–3), 285–293.

Wikipedia. (2016). Middle East—Wikipedia, the free encyclopedia. Retrieved from https:// en.wikipedia.org/wiki/Middle_East. 2016-09-22

Woodhouse, D. (2008). University rankings meaningless. Retrieved from http://www.university worldnews.com/article.php?story=20080904152335140. 2016-11-06

Zayed University. (2014). Sheikh Mohammed Bin Zayed higher education grant. Retrieved from http://www.zu.ac.ae/main/en/alumni/con_education/smbzheg_2014.aspx

Chapter 7

QS World University Rankings

Ben Sowter, David Reggio, and Shadi Hijazi

QS Intelligence Unit, London, United Kingdom

Contents

7.1 Introduction

7.1.1 Rankings and Their Flaws

Choosing an institution has increasingly become a strategic move for both students and scholars: students aim for the strongest outcome, which today is pitched as employment; scholars aim for an environment cultivating academic and social impact. The decision has been crafted by a drive for increased internationalization, global engagement, and innovation over the last few decades. Amplified demands for student outcomes, transferable research, and solution building have also seen an increase in the number of alternative study destinations. As a result, established names and historical prestige are no longer the only aspects to an institution's "gravitational pull" in a world of shared initiatives, geopolitical change, and scientific development. New epochal alternatives and research destinations have emerged in countries that have not traditionally been considered destinations for education, research, and contemporary industries.

Consistent data and qualitative insight are two pillars supporting the election of an informed choice in a market that is quasi-saturated by marketing messages and information. A student-centered approach signals that a metric-driven system must be mindful of the beneficiaries and stakeholders that it seeks to address and inform and that data be sourced, treated, and presented in a way that is meaningful, pertinent to diverse needs and aspirations, and useable across key sectors. Since its foundation in 1990, QS established itself as one of the world's leading specialists in higher education as well as a career information and solution provider. Its services include student recruitment events in 45 countries, publications, and career guides, leading websites such as TopMBA.com and TopUniversities.com and research and software solutions in the higher education sector. The global nature and progressive initiatives hallmarking the evolution of QS stem from its core mission philosophy, namely, "to enable motivated people around the world to fulfil their potential by fostering international mobility, educational achievement and career development."

Within this context, QS World University Rankings was launched to meet a contemporary demand for insight delivered through a structured and transparent process, where information is graspable and immediate and where the realities defining academic institutions around the world, and higher education, have been captured by numerical measure. At the same time, rankings need to be sophisticated enough to encompass excellence in a globally comparable and equitable manner. This balance between simplicity, necessary for the public, and a level of sophistication that is able to capture the research realities of an institution is the central challenge facing any method of measurement. It is recognized that all methodologies have their limits, and QS Rankings is no exception to the scientific and social thirst for refinement and for building an evidence base that is actionable by a variety of stakeholders. Furthermore, rankings are not a stand-alone beacon on this higher education horizon, and there are other complimentary evaluation and insight systems available: quality assurance (QA) measures created by QA agencies and accountability measures instituted by governments are two such examples (Shin et al., 2011). Considering the system user, however, it is ranking that has arguably the strength and appeal of familiarity. Unlike QA reports and ministerial policy, rankings remain the most accessible of performance systems available to the general public. Ranking is defined as the practice of listing universities in an ordered list based on performance indicators (Dill, 2009), and this signals its usability and accessibility by policy makers and students alike. For the former, it is a tool to fashion a higher education mandate which qualifies funding and national university standards, exemplified by the Russian Federation's 5-100 project. For the latter, focus groups have revealed that students value rankings as part of a decision-making mix rather than popular opinion (Karzunina et al., 2015).

7.1.2 Rankings and Ratings

To meet the need for refinement and adaptability to evolving institutional horizons, more comprehensive initiatives that include multidimensional assessments of a university's performance have been introduced. These include the EU project U-Multirank and the QS Stars rating system.

U-Multirank started as an alternative to rankings, professing a "different approach to the existing global rankings of universities [where] U-Multirank does not produce a combined, weighted score across these different areas of performance and then use these scores to produce a numbered league table of the world's 'top' 100 universities" (Multirank, 2016). U-Multirank, aware of the need for user intelligence, introduced two example sets of rankings platforms supported by a rich database: *Research and Research Linkages Ranking* and *Teaching & Learning Rankings*.

Multidimensional rating systems have their clear advantages, particularly when conducted in a robust fashion with proper validation and cooperation mechanisms. Global rankings, however, need to be more inclusive, requiring a careful, rigorous choice of structured data that can be verified at scale. The QS Stars rating system follows the same logic but adopts a different approach to multidimensional assessment. QS Stars is based on a comprehensive methodology that provides a wider range of categories for universities to be assessed on, enabling recognition of distinctive strengths be they in teaching, cultural initiatives, research, and innovation. The virtue of such a system is in unifying an institution's academic focus with the specific requirements of students. Universities are in turn assessed against a set standard in a series of compulsory and elective categories best fitting to strategy and strengths.

Rating systems require a thorough validation of data and close collaboration with an institution. The process of data submission and validation requires a period of joint work between a dedicated QS Intelligence Unit analyst and university administration: the university provides QS with evidence pertinent to the indicators which is in turn verified according to set parameters and quality control prior to acceptance. As an end result, the university is awarded an overall Stars rating as well as a rating for each category. Because of the nature of collaboration between QS and the university, the categories considered in a Stars audit can be more inclusive than what can be measured in more global ranking exercises.

As such, a wider more comprehensive and granular picture based on accurate, validated, and contemporary data can be gathered about the institutions where students are thus enabled to compare audited universities through the QS website http://www.topuniversities.com/qs-stars.

The QS Stars categories are currently set out in the following manner (QS Top Universities, 2016):

1. *Research*: Indicators considered here include assessments of research quality among academics, productivity (i.e., number of papers published), citations (i.e., how recognized and referred to those papers are by other academics), and awards (e.g., Nobel Prizes or Fields Medals).

2. *Teaching*: Indicators used in teaching quality assessments are student feedback through national student surveys, further study rate, and student–faculty ratio.

3. *Employability*: Indicators in this area include surveys of employers, graduate employment rates, and career service support.

4. *Internationalization*: Indicators include the proportion of international students and staff, the numbers of exchange students arriving and departing, the number of nationalities represented in the student body, the number and strength of international partnerships with other universities, and the presence of religious facilities.

5. *Facilities*: Indicators regarding the university's infrastructure such as sporting, IT, library, and medical facilities, as well as the number of student societies, are considered within this criterion.

6. *Online/distance learning*: This category is more suited for online institutions and examines various indicators such as student services and technology, track record, student–faculty engagement, student interaction, commitment to online learning, and reputation of the university.

7. *Social responsibility*: This measures the university's engagement with the local community through investment as well as in charity work and disaster relief. It also includes regional human capital development and environmental awareness.

8. *Innovation*: Indicators here focus on the university's output in terms of registered patents, spinoff companies, and industrial research.

9. *Arts and culture*: Indicators include the number of concerts and exhibitions organized by the institution, the number of credits, and cultural awards and cultural investment.

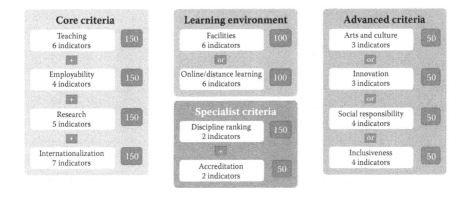

Figure 7.1   Indicators considered in QS Stars methodology.

10. *Inclusiveness*: Indicators are concerned with the accessibility of the university to students, particularly at scholarships and bursaries, disability access, gender balance, and low-income outreach.
11. *Specialist criteria*: These are focused on assessing the university's excellence in a narrow field and looks at accreditations and discipline rankings.

More details about QS Stars methodology and the weights of the indicators are available at the QS Intelligence Unit website, http://www.iu.qs.com/services/qs-stars/qs -stars-methodology, and an overview of the methodology is presented in Figure 7.1.

7.2 Pillars in QS World University Rankings

As global rankings are broad in scope, application, and representation, the number of measures used in their construction must be scalable and verifiable. These measures at the same time must capture the most relevant dimensions for beneficiaries, the most important of which are students. This said, a completely comprehensive and objective perspective is ideal in an aggregate ranking, sooner than the outcome: whereas it is not unusual for an institution to track progress and performance through hundreds of metrics, global rankings are selective and limit the measures tracked. Hazelkorn's (2011) review of global rankings classified the various indicators and weightings used by international rankings into research (including publications, citations, innovation and research income generation, teaching [including employability], society, and international impact (pp. 32–39). In this regard, the choice of ranking indicators and their weights becomes of particular importance. As the selection of indicators and the weightings applied will always introduce an element of bias, it is equally important for users of this qualitative system to examine what is being measured and ensure that it is compatible with their requirements. QS, in turn, has identified four main pillars that contribute to a world-class university. These are (1) research, (2) teaching, (3) employability, and (4) internationalization.

7.2.1 Research

Institutions are today committed to achieving excellence in research and scholarship and to ensuring that research contributes to the well-being of society, providing a creative and supportive environment in which ideas are generated and applied. The scale of today's research questions, and the equipment required to study them, demands that researchers work collaboratively across national boundaries and disciplines. It is crucial that in these emerged and emerging research environments that institutions sustain the best research through international engagement at every level from collaborations between individual academics to multidimensional institutional agreements.

Research thus continues to be a dominant indicator for measurement in most university rankings, and world university rankings almost exclusively favor research-intensive institutions because of the ever-expanding horizon of science and its evolving relation with modern society. The ARWU includes more than 1,200 universities in terms of recognized excellence. Measures of such excellence include those indicative of research quality such as Nobel Prizes and Fields Medals awarded among academic staff and/or alumni or academic output such as Highly Cited Researchers or papers published in *Nature* or *Science*. There are also quantitative measures: papers indexed by *Science Citation Index Expanded* (SCIE) and the *Social Science Citation Index* (SSCI). These indicators combined account for 90% of the score, with the remaining 10% introducing the number of full-time equivalent (FTE) academic staff when available (ARWU, 2016).

The research element in QS Rankings constitutes the highest contribution to the overall results and is the metric that has undergone progressive methodological refinement. The QS approach to measuring research in the world rankings uses two indicators to assess an institution's research contribution: (1) citations per faculty (worth 20%), based on bibliographic data for research output and the number of verified FTE academic staff in academic institutions included in published rankings; and (2) academic reputation (40%), measured by peer reviews from international academics through the QS Global Academic Survey.

7.2.2 Teaching

A university is a working reality of high complexity and various dimensions. The challenge therefore is as to what precisely is to be measured in order for quality to be ascertained correctly. There are various possible proxies for measuring the teaching and learning environment at institutions, some of which are included in multidimensional ratings, but there is no practical way of assessing teaching quality itself. The

Organization for Economic Cooperation and Development (OECD), for example, has looked at the feasibility of measuring teaching quality from a testing standpoint through a proposed Assessment of Higher Education Learning Outcomes. The results of the feasibility study revealed that assessing learning outcomes is a costly exercise (OECD, 2014) and that learning outcomes remain all but a proxy for teaching quality.

Surveying students is another alternative, yet one which is insufficient and open to bias (either by culture or by association): most students do not have a comparative frame of reference, and they have an invested stake in the outcome. Cultural mind-sets and established values can potentially affect the underlying theoretical concept where it is matched to the categories present in Likert-style surveys (Hui & Triandis, 1989). Collectivist countries, moreover, appreciate harmony and avoid confrontation, as a result, extreme response choices are bypassed in favor of answers which will not rock the boat or contradict established values. Contrastingly, respondents from countries whose cultures, practices, and consumer tendencies are built around more individualist identities prefer direct, unobstructed answers (Hui & Triandis, 1989; Dolnicar & Grun, 2007; Harzing et al., 2012). Student surveys, rolled out on a global level to yield comparison across different institutions, are therefore limited by bias and, more broadly, the sociohistorical determinants fashioning a culture and its practices.

The QS approach to measuring teaching in the world rankings uses the student–faculty ratio indicator that contributes 20% to the overall ranking of the university. It is the only globally comparable and available indicator that has been identified to address the stated objective of evaluating teaching quality. Data used in this indicator are collected and verified by QS data teams.

7.2.3 Internationalization

Higher education is today characterized by the drive toward international mobility and what is broadly deemed *global engagement*. While the former has always featured, to various extents and forms, as an institutional practice, the latter concerns global citizenship and specific strategic mechanisms facilitating research partnerships, training, and experiences for students and graduates as well as the internationalization of all curricula. Measuring the level of internationalization at an institution has received comprehensive treatment in the QS Stars rating system encompassing numbers, diversity, short-term internationalization, international collaboration, and the availability of religious facilities. In a global ranking context, however, only two indicators are appropriate in terms of their suitability and practicality of data collection: (1) the ratio of international students and (2) the ratio of international faculty. The ratio of international faculty is introduced as a percentage of international academic staff, which is, more broadly speaking, both an indication of diversity in an academic environment and its appeal to international scholars. Combined, the international student ratio and international faculty ratio contribute to 10% of the overall ranking of an institution. Data used in this indicator are also collected and verified by QS data teams.

7.2.4 Employability

As a student-oriented ranking organization, QS has always considered employability as an important element in the decision-making process of students. This fact informs the portfolio of QS activities around the world and QS global focus group research with students. It is invariably agreed by students attending QS fairs and participating in focus groups that employability is vital. Students thus value subject rankings over holistic rankings (Karzunina et al., 2015).

A common approach to the evaluation of employability in domestic rankings, and in QS Stars, is the graduate employment rate. This, however, would not work in global rankings as the top universities in the world naturally have a high level of employment and historically established links with industry. The demographic and economic situations of various countries therefore introduce a strong bias, one not related to the university's performance.

Accounting for 10% of the QS World University Rankings, the employer reputation indicator is unique among the current international evaluations in that it canvasses employer opinion on the quality of an institution's graduates.

7.3 Data Collection and Validation

Data needed to compile the world university rankings are drawn from three sources: (1) the QS Intelligence Unit data acquisition team, (2) the Scopus database, and (3) QS Academic and Employer Surveys.

7.3.1 Data Acquisition Team

The Quacquarelli Symonds Intelligence Unit (QSIU) data acquisition team is responsible for collaborating with universities in the data submission, validation, and updating processes. These cycles are defined by key tasks, in addition to collecting updated and validated measurements for the relevant measurements. For world university rankings, the data acquisition team is responsible for providing data that feed into the FTE number of students, faculty members, international students, and international faculty. There are other data that universities involved in regional and specialized rankings need also to supply, including numbers of international exchange students to measure shorter mobility phases. Faculty numbers used in the rankings are totals of both teaching and research staff. Separating these two elements would be ideal but is not as of yet possible.

QS introduced a custom-built database system called QS Core in 2010, which is the main platform of collaboration between the Intelligence Unit and universities. Notably, the platform facilitates institutional users to have more control over their data submission, which has yielded increasing accuracy and greater transparency.

All communication regarding the validation or inquirers flows through the same system. A complete list of the data that can be submitted by the university to the core system and the exact definition of each of these data points can be downloaded from http://www.iu.qs.com/data-services/definitions.

A partial list of the 20,000 universities in the world (an approximate number) was introduced in 2004. This initial list considered the world's top 500 universities in the citations per paper index, and since that time, four streams have come into play to facilitate inclusion:

1. *Domestic ranking performance*: The data acquisition team in QS tracks a growing number of domestic rankings to make sure that well-recognized institutions in their countries are included.
2. *Survey performance*: Respondents to the Academic and Employer Reputation Surveys run by QSIU are invited to suggest any institutions that they feel may have been omitted from the exercise.
3. *Geographical balancing*: As a global ranking, the balance in the number of institutions from given countries and regions is periodically reviewed.
4. *Direct case submission*: When institutions approach QS directly to request inclusion, QSIU evaluates each case on its merits drawing comparison against institutions already included in the rankings. Subject to certain prerequisites and performance indicators being met, additional institutions are included.

The data acquisition team is responsible for validating data used for the classification of academic institutions that help users of QS Rankings focus on institutional comparison. QS tables introduced the classifications in 2009, and they are consistent with the Berlin Principles on Rankings where any comparative exercise ought to take into account the different typologies of the institutions, and labels of such typologies are to be present in the aggregate ranking list (IREG, 2006). Classifications in QS Rankings are simple and loosely based on the Carnegie Classification of Institutions of Higher Education in the United States (available online at http://carnegieclassifications.iu.edu), but operated on a much simpler basis.

QS classifications take into account the following key aspects of each university to assign their label:

1. *Size*: Size is based on the (FTE) size of the degree-seeking student body.
2. *Subject range*: There are four categories based on the institution's provision of programs in the five broad faculty areas used in the university rankings. Due to radically different publication habits and patterns in medicine, an additional category is added based on whether the subject institution has a medical school.

3. *Age*: Five age bands are based on supplied foundation years.
4. *Research intensity*: Research intensity is based on the number of documents annually retrieved from Scopus in the 5-year period preceding the application of the classification. The thresholds required to reach the different levels are different dependent on the institution classification on size and subject range.

7.3.2 Bibliometrics

Citations, relative to the size of the institution and evaluated as such, are the best understood and most widely accepted measure of research strength in quantity and quality. QS World University Rankings has adopted a citation *per faculty member* approach since its inception in 2004.

This element to the rankings has undergone the greatest methodological refinement. In the first three editions of the QS World University Rankings, data from the Essential Science Indicators, a subset of the WoS, were used for publication and citation data. This was, however, lacking in terms of international coverage of indexed journals. Therefore, in 2007, citations and paper data for rankings were collected using Scopus database. This resulted in an immediate improvement in the number of institutions considered in citation analysis and the ranking exercise as a whole.

Scopus is the world's largest abstract and citation database of peer-reviewed research literature. QS receives a snapshot of the Scopus database from Elsevier and maps it to its own internal systems to match bibliographic data to institutional and survey data. In the most recent round of rankings, 6,805 Scopus IDs were mapped to 2,557 QS Core IDs. Scopus is a rapidly evolving system, and thus, the QS database may differ significantly from the latest versions of Scopus online.

In addition to the number of papers and citations, *h*-index as a subject-level indicator for the quality of research is another Bibliometric measurement applied in the QS subject rankings. The *h*-index, developed by J. E. Hirsch, qualifies the impact and quantity of individual author research output. Hirsch (2005) defined the *h*-index as follows: "A scientist has index *h* if *h* of his or her N_p papers have at least *h* citations each and the other $(N_p - h)$ papers have $\leq h$ citations each."

7.3.3 Surveys

QS invests strongly in the software, design, effective communication, QA, and database management of its surveys. The integrity of survey responses holds paramount importance where QS runs a sophisticated screening analysis to detect anomalous responses, routinely discarding invalid and suspicious responses. Sources have to be verified as being completely independent and unsolicited; therefore, any attempt to manipulate results or influence nomination can result in the disqualification of all responses for an institution that year. While the cooperation of the universities is appreciated, it is also not permitted to solicit specific responses from expected respondents to any survey contributing to QS Rankings.

Two surveys contribute to 50% of the overall ranking score: (1) The academic reputation indicator score is based on the QS Global Academic Survey, the largest survey of its type on the opinions of academics globally with a response of almost 75,000, reflecting the strong awareness and interest in QS and the rankings. To boost the size and stability of the sample, QS combines responses from the last 5 years, where only the most recent response of every respondent is counted. Results are globally sourced in terms of respondent locations and areas of knowledge. The consequent findings are made available to the public via the QSIU website http://www.iu.qs.com/academic-survey-responses. (2) The QS Employer Survey has been running since 1990 and contributes to a number of key research initiatives operated by the QS Intelligence Unit. In the 2016/2017 rankings, 37,871 responses were used, compared to only 17,000 in 2011, reflecting a strong growth in interest, particularly in Latin America, China, India, and Russia. Country-level data are publically available at http://www.iu.qs.com/employer-survey-responses/.

7.4 Number Crunching

Further calculations and modifications are consequently performed on the collected data to arrive at an overall rank. Some of these processes were introduced at a later stage to improve the accuracy and consistency of the rankings.

7.4.1 Bibliometrics

Citation and paper counts used in the rankings consider the university's output in the previous 5 years. In 2011, self-citations were excluded from the analysis to improve performance in bibliographic research metrics. Three further modifications were introduced in 2015, namely, the following:

1. *Affiliation caps for papers*: Highly cited material produced by very large research groups gives substantial credit to institutions that have only contributed in very small part to the work—it is not unusual that research papers can carry a large number of authors and affiliations. As such, these were excluded from the count in 2015. Originally, papers featuring authors from more than 10 affiliated institutions were excluded. This was replaced in 2016 by a variable cap calibrated to ensure that no more than 0.1% of research is excluded from any discipline. This is more sensitive to the publishing patterns of a discipline to which a given paper belongs: less papers are consequently now excluded from subjects such as physics, earth and marine science, and the biological sciences, where it is more common for larger research groups to collaborate to a single publication.
2. *Removal of superfluous research content types*: The focus on high-quality research signaled that certain content types had to be removed from Scopus

data. This improvement exercise was jointly initiated in consultation with Elsevier and resulted in the removal of the following: abstract reports, conference reviews, editorials, errata letters, notes, press releases, and short surveys.

3. *Faculty area normalization*: Due to publishing patterns, a straight ratio of citations per faculty places a strong emphasis on life sciences and medicine. After a lengthy process of consultation with advisors and participants, QS sought to adopt a model equalizing the influence of research in five key faculty areas. The effects of this faculty area normalization are clarified in Figure 7.2. The calculation of normalized total citation count (NTCC) is shown in Figure 7.3 with technical explanations available online (QSIU, 2015).

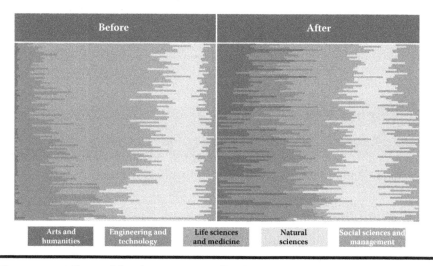

Figure 7.2   Effects of faculty area normalization.

$$NTCC \equiv \frac{n}{n_{fa}} \sum_{f=1}^{5} c_f w_f a_f$$

$$w_f \equiv \frac{n_{fa}}{5\,x_f}$$

$$a_{ah,ss} \equiv \frac{r_f - r_{fmin}}{1 - r_{fmin}}$$

$$r_{ah,ss} \equiv min\left\{ \frac{p_f}{p_{fmax}} \middle| 1 \right\}$$

$$a_{et,ls,ns} \equiv \frac{5 - (a_{ah} + a_{ss})}{3}$$

n = total citation count prior to normalization
n_{fa} = sum of total citation count across the five faculty areas
c_f = count of citations for the given faculty area for the institution in question
w_f = weighting factor for the given faculty area
a_f = weighting adjustment for the given faculty area
x_f = count of citations for the given faculty area
r_f = ratio of a country's papers in the faculty area to the most productive country in the faculty area, in relative terms
r_{fmin} = the lowest value of r_f across all countries
f = current faculty area, which can be one of:
   ah = Arts and humanities
   et = Engineering and technology
   ls = Life sciences and medicine
   ns = Natural sciences
   ss = Social sciences and management
p_f = mean proportion of papers from the faculty area for the institution's home country
p_{fmax} = the maximum value of p_f where the paper count in that faculty area for the given country exceeds the global average

Figure 7.3   Calculation of the normalized total citation count (NTCC).

7.4.2 Surveys

Collection of robust data in the two surveys administered by QS is only the starting point. The analysis to reach a score in the academic reputation and employer reputation indicators involve further processing. For each of the two surveys, the following steps are followed for each of the five subject areas:

1. Devising weightings based on the regions with which respondents consider themselves familiar.
2. Deriving a weighted count of international respondents in favor of each institution ensuring that any self-references are excluded.
3. Deriving a count of domestic respondents in favor of each institution adjusted against the number of institutions available for selection in that country and the total response from that country ensuring that any self-references are excluded.
4. Applying a straight scaling to each of these to achieve a score out of 100.
5. Combining the two scores with the following weightings: 85% international, 15% domestic (for the academic survey), 70% international, and 30% domestic (for the employer survey). These numbers are based on the analysis of responses received before separation of domestic and international responses.
6. Squaring the result to draw in the outliers as excellence in one of our five areas should have an influence, but not too much of influence.
7. Scaling the rooted score to present a score out of 100 for the given faculty area.

Once these are completed, the five totals of the faculty areas are combined with equal weighting to result in a final score from each survey, which will then be standardized, relative to the sample of institutions. As a result of this process, final scores for the academic reputation indicator and the employer reputation indicator are reached and used in the aggregate rankings.

7.4.3 Damping

Damping, or smoothing, was introduced in 2010 to spread sudden changes occurring in 1 year over a longer period in order to avoid sudden reactions that might be a result of policy change or significant changes to a university's strategy or conditions. The details of this mechanism and the thresholds used vary depending on the indicator, and thresholds can be absolute or relative to previous years. Such a mechanism ensures sustainability and consistency of performance as well as increased stability and reliability of the rankings.

7.4.4 Overall Scores and Ranks

Unifying ranking indicators into a singular composite score is the final step. Normalization, or standardization, is necessary here to make the various data collected compatible and to apply the weightings. Originally, the approach was simple, including finding the top-scoring institutions in every indicator, awarding these 100 notional points and scaling the remaining entries proportionally to that top performer. Since 2007, a z-score has been utilized where the resulting z-scores are scaled between 1 and 100 for each indicator, weights are applied, and the total score is reached. Institutions are subsequently ranked based on this final score.

7.5 Beyond World Rankings

Through our work with students and institutions, it became evident that further measurements are needed to fully grasp the advances and trends of today's higher education horizon, thereby providing further insight and accuracy. In this regard, three different approaches have emerged: (1) geographical and regional distinction through regional rankings, (2) specialized subject rankings, and (3) initiatives to measure excellence in a focused manner such as the Reimagine Education Awards, QS Employability Rankings, and public engagement data-driven initiatives such as the QS EduData Summit. In each of these approaches, methods that are not available or suitable on a global scale were adopted, and further refinement was possible. Presently, QS publishes five regional rankings (Asia; Latin America; BRICS countries [Brazil, Russia, India, China, and South Africa]; Emerging Europe and Central Asia; and the Arab region). Although the approach and intention are similar for each of these, weights and new indicators have been introduced as specific characteristics pertinent to particular regions. New indicators include inbound and outbound exchange students, staff with PhDs, and web impact and international research networks. Furthermore, the citation per faculty indicator was replaced by two separate indicators: (1) paper per faculty to assess quantitative research output and (2) citations per papers as an indicator of research quality. All indicators and weights used in the various QS Rankings are shown in Figure 7.4.

Subject ranking is another area that has received keen interest from all global regions, stakeholders, and beneficiaries. The QS World University Rankings by Subject 2016 heralded a new standard of insight in such ranking indices by expanding the number of subjects covered to 42. Another advance was the inclusion of employer reputation into the world rankings. Further signs of development have been evidenced by the QS Graduate Employability Rankings, the research process of which begun in 2014. At this stage, the exercise remains a pilot and universities have only been ranked if they elected to actively participate in QS's data collection processes or if enough publicly available data were available to permit their inclusion. The development of the methodology, and the latest published tables of this new frontier in QS Rankings, is available online (QSIU, 2016).

Behind the *simple* number of a ranked institution, there is a highly dynamic and evolving process. The QS team remains on a continuous quest to identify gaps and seek further data and methodological refinement in order to improve the accuracy

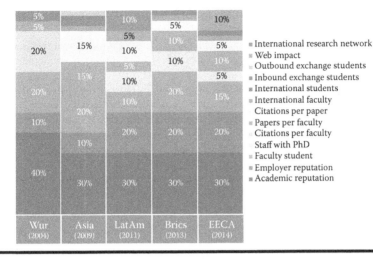

Figure 7.4   Indicators and weights used in various QS Rankings.

of rankings and other regional and specialized tables. In all of this, QS is committed to its partnership with the world's academic institutions, with a level of engagement that involves workshops, methodology consultations, and prerelease institutional profiles. Transparency, continued accuracy, and relevance to an ever-changing horizon are the hallmarks to the identity of QS Rankings, which continue to be a powerful tool for students as well as all stakeholders.

References

ARWU (Academic Rankings of World Universities). (2016). ARWU 2016 methodology. Retrieved October 5, 2016, from http://www.shanghairanking.com/ARWU -Methodology-2016.html

Dill, D. D. (2009). Convergence and diversity: The role and influence of university ranking. In M. K. Barbara & S. Bjorn (Eds.), *University Rankings, Diversity, and the New Landscape of Higher Education* (pp. 97–116). Rotterdam, Netherlands: Sense Publishers.

Dolnicar, S., & Grun, B. (2007). Cross-cultural differences in survey response patterns. *International Marketing Review*, 24 (2), 127–143.

Harzing, A. W., Brown, M., Köster, K., & Zhao, S. (2012). Response style differences in cross-national research: Dispositional and situational determinants. *Management International Review*, 52 (3), 341–363.

Hazelkorn, E. (2011). *Rankings and the Reshaping of Higher Education: The Battle for World-Class Excellence*. Hampshire, New York: Palgrave Macmillan.

Hirsch, J. E. (2005). An index to quantify an individual's scientific research output. *Proceedings of the National Academy of Sciences of the United States of America*, 102 (46), 16569–16572.

Hui, C. H., & Triandis, H. C. (1989). Effects of culture and response format on extreme response style. *Journal of Cross-Cultural Psychology*, 20 (3), 296–309.

IREG (International Ranking Expert Group). (2006). Berlin Principles on Ranking of higher education institutions. Retrieved October 6, 2016, from http://ireg-observatory.org /en/berlin-principles

Karzunina, D., Bridgestock, B., & Philippou, G. (2015). How Do Students Use Rankings? The role of university rankings in international student choice. London: QSIU available online at http://www.iu.qs.com/product/how-do-students-use-rankings/

Multirank. (2016). Our approach to ranking. Retrieved October 4, 2016, from http://www .umultirank.org/#!/about/methodology/approach-to-ranking

OECD. (2014). Testing student and university performance globally: OECD's AHELO. Retrieved October 6, 2016, from http://www.oecd.org/edu/skills-beyond-school /testingstudentanduniversityperformancegloballyoecdsahelo.htm

QSIU. (2015). Faculty area normalization—Technical explanation—QS. Retrieved October 7, 2016, from http://www.topuniversities.com/qs-stars/qs-stars-methodology

QSIU. (2016). QS Graduate Employability Rankings. Retrieved October 7, 2016, from http://www.topuniversities.com/qs-stars/qs-stars-methodology

QS Top Universities. (2016). QS Stars university ratings. Retrieved October 4, 2016, from http://www.topuniversities.com/qs-stars/qs-stars-methodology

Shin, J. C., Toutkoushian, R. K., & Teichler, U. (2011). *University Rankings: Theoretical Basis, Methodology and Impacts on Global Higher Education*. Dordrecht, Netherlands: Springer.

Chapter 8

Times Higher Education World University Rankings

Duncan Ross

Times Higher Education, TES Global Ltd, London, United Kingdom

Contents

Times Higher Education (THE) was founded in 1971 as a supplement to the *Times* newspaper, as a journal focused on the world of higher education—a sister publication to the Times Education Supplement. Since 2005, we have been independent of the *Times*, but the journalistic background and readership of our business has given us a unique perspective on the issue of performance and comparison in world higher education.

The first THE World University Rankings was published in conjunction with QS in 2004. In 2010, THE developed its own methodology, with a wider set of indicators, and in 2015, we moved the creation and ownership of the ranking entirely in-house.

In that same time period, the ranking has developed its methodology and has extended its reach. Starting with just 200 universities, largely from the United

States and Western Europe, it now reflects the performance of 980 universities from 97 countries.

In this chapter, I will look (briefly) at our methodology and explore a few key aspects that are specific to the issues around measurement: quality, normalization, and metric complexity. Finally, I will address some of the potential future developments of rankings.

8.1 THE World University Rankings

This section will give an overview of our methodology (for more detailed descriptions, please refer to our website and the audit performed by PricewaterhouseCoopers).

Our ranking unashamedly focuses on research-intensive universities with a broad curriculum and an undergraduate teaching role. Although every university in the world is free to provide data that might result in them being ranked, we limit our scope using the following criteria:

■ More than 1,000 qualifying publications over a 5-year period
■ Undergraduate teaching
■ A broad curriculum

For most universities, a section of data is provided directly*—we invite universities to submit data to our portal, both at an overall university level and by eight broad subjects.

These data provide the first of our three data sources.

Our second data source is our annual Academic Reputation Survey. This survey is carried out through a random sampling of academics selected from the Scopus database. These are then balanced by country—our survey has been designed to reflect worldwide opinion.

The final dataset is the Scopus database itself. From this, we take information on the number of citations, the academic productivity, and the degree of international collaboration for each university.

These data are used to create 13 separate metrics, which are, in turn, grouped into 5 broad areas: teaching, research, citations, industry income, and international outlook (Figure 8.1).

However, in order to be able to sensibly compare and combine these variables, we need to find appropriate normalization methods. These break down into three areas:

■ Size
■ Subject
■ Distribution

---

* The exceptions are universities from the United Kingdom, whose data are sourced from the Higher Education Statistics Agency.

Figure 8.1 **THE methodology (also available at http://duncan3ross.github.io/WUR/).**

Our first goal is to prevent a situation where the largest institutions always win. To do this, many of our variables are represented as ratios: either to the size of the student body or to the size of the faculty.

Our second goal is to account for the very different attributes relative to different subjects. An early contention for rankings was that they tended to reward institutions that specialized in medicine or science. By treating the data by subject, and then combining the results, we avoid some of the worst aspects of this.

Our final goal is to ensure that we can account for the distributions of the variables and combine them sensibly. This is performed using z-score normalization, except in the case of the reputation survey data, where an exponential approach is better suited to the data.

The final step of creating the ranking is to create the overall score. In our methodology, each of the 13 metrics is assigned a weighting. These weightings have been consistent since the second version of our approach was released in 2010.

Although I have focused on the World University Rankings, THE also produces a number of subsidiary rankings: for the broad subject areas and for geographical regions. For these rankings, we are able to adjust the proportions of each variable to more accurately reflect the issues and concerns of that region or subject.

8.2 Notable Scientometric Issues

Of course, before starting this section, it is important to emphasize that rankings themselves are not a science. We are, in effect, generating a balanced scorecard that (we hope) gives an overall estimation of the performance of universities. If you disagree with our assessment, then there is nothing preventing you from voicing that disagreement. We also acknowledge that there are many aspects of university performance that cannot be adequately covered, especially in international rankings.

These are not, however, the challenges that particularly interest me. Rather, the challenges are these: data quality, comparison, and metric complexity.

8.2.1 Data Quality

In the absence of an internationally accepted and open dataset on universities, we collect data ourselves. Ensuring that the quality is appropriate* takes a great deal of time and is inevitably an area of continuous improvement.

When we gather data, we are able to validate it against previous years' submissions and to identify and flag potential challenges to the universities submitting the data. We also manually check the data for outliers, which are followed up directly with universities.

---

* Data quality should never be assumed to be good.

In addition, we find that universities themselves act as validators of each other's input. When a university scores especially highly in a particular category, we find that there is no shortage of academics from other institutions that are willing to challenge the data. We treat these challenges seriously, and will investigate, and if necessary take action.

Elsevier also takes great care to curate data in the Scopus database on citations appropriately. We work closely with Elsevier to make sure that the attribution of papers to institutions is correct—this is probably the most challenging data quality aspect of the Bibliometric dataset. The Scopus committee also takes measures to ensure that the journals that they index are genuine academic journals. They can, and will, suspend journals from the dataset.

8.2.2 Comparison

When we want to compare variables, we need to convert them into a value where that comparison makes sense. This is why we use z-scores.

Effectively, we are converting our metrics into an estimation of the performance within a distribution and then comparing these performances. However, because these distributions are based on the dataset, they will change as the number of universities (and indeed the scores of the universities) changes (Figure 8.2).

In particular, as we increased the scope of the World University Rankings, from 400 universities in 2014–2015 to 800 in 2015–2016, and now to 980 in 2016–2017, the distributions for each metric and pillar altered. For this reason, although we believe that there is a great deal of insight available from investigating changes from year to year, any such analysis requires a degree of caution.

8.2.3 Metric Complexity

There is a well-known dilemma in the data science world: how to balance between the subtlety of a measure and its interpretability. This is important because, on the one hand, we want to make sure that our metrics are not obviously mismeasuring and, on the other hand, we want to make sure that they are easily understood.

This is particularly important when we look at metrics around research performance. Ignoring, for a moment, some of the challenges around kilo-author papers, there are five metrics in our rankings to which we apply a subject weighting.

These are the following:

- Doctorates to academic staff
- Papers to academic staff
- Research income to academic staff
- Field-weighted citations
- Papers with one or more international author

Metric	%	Relative	Norm	Subject weight	PPP
Doctorates to academic staff	6	Staff	Z	Yes	
Doctorates to bachelor's degrees awarded	2.25	Student body	Z		
Teaching reputation	15		Exp		
Income to academic staff	2.25	Staff	Z		Yes
Staff-to-student ratio	4.5	Staff	Z		
Papers to academic staff	6	Staff	Z	Yes	
Research income to academic staff	6	Staff	Z	Yes	Yes
Research reputation	18		Exp		
Field-weighted citations	30	Papers	Z	Yes*	
Industry research income to academic staff	2.5	Staff	Z		Yes
International to domestic staff	2.5	Staff	Z		
International to domestic student	2.5	Student body	Z		
Publications with international author	2.5		Z	Yes	

THE Data Points

Figure 8.2 Normalization approaches.

The reason for applying the subject weighting is simple: different subjects have very different traditions of publication and funding. Without this weighting, we would effectively be measuring the subject mix of the university rather than its performance.

In most of the cases, we normalize using the eight broad subject categories, but in the case of field-weighted citations, we use a much finer level of detail. Working with our partner Elsevier, we first identify the average number of citations expected within each of 334 separate academic fields and then compare each paper to that expected value. Once we have done that, we create an average of this value across a university's output. These values are then normalized across the cohort of universities.

The challenge, of course, then comes with interpretation. A value will change between years, as the papers included change and as the number of universities in the cohort changes. In addition, the exclusion of journals that are suppressed by the Scopus committee can significantly impact results for specific universities.

An obvious question is around the level to which we normalize. Is it too detailed? Is it too coarse? In the case of field-weighted citations, I believe that we have struck the correct balance: the metric is complex enough to capture the majority of the differences in publication approaches, without being so complex that it becomes unstable.

8.3 Future Ranking Directions

When introducing our methodology, I made it clear that our focus has been leading research-led institutions. There are very good reasons why this has been the case, not least around the availability of internationally comparable data on research.

However, research is only one mission of higher education. Another of the key missions is teaching. Together with the *Wall Street Journal*, THE has just released its first U.S. College Ranking.

This ranking is different. Rather than looking at teaching as only one of the areas of excellence, we have instead focused primarily on teaching. Although there are measurements that reflect research, these are very much present in support of teaching, rather than the other way around.

As you can see from the methodology diagram, rather than the five areas of the World University Rankings, we focus on four teaching-related areas:

■ Resources
■ Engagement
■ Outcomes
■ Environment

This allows us to explore teaching effectively for the first time, based on a sound theoretical background.

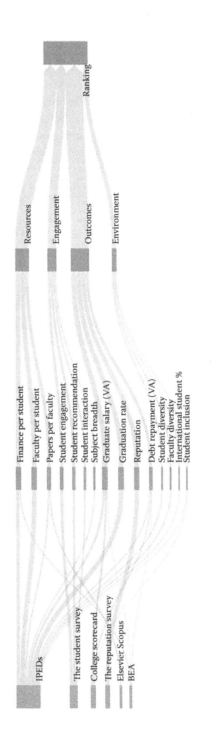

Figure 8.3 WSJ/THE College Rankings methodology.

When we do this, we also include the voice of the student—something that is usually missing from university rankings. And this provides useful insight—part of which is that, in the United States at least, universities that excel at research are not necessarily good at engaging with their students (Figure 8.3).

Another direction that is often suggested is for more detailed subject rankings. This presents some difficulties. Because we rely on institutions to provide data, we are creating a significant burden on the universities to provide the data to our specifications. Each additional data point that we request requires additional work.

We could, of course, simply rely on externally sourced data to do this. But, anecdotally, we have been told of universities ranked highly in subjects that they do not and have never taught. This is not satisfactory.

There is, however, undoubtedly a demand for this kind of information, so it is likely that we will try to extend in this direction.

Finally, there are bespoke rankings around factors such as innovation, social mobility, and more. I think that these areas offer real opportunities for giving a wider set of universities the opportunity to shine.

Chapter 9

Scimago Institutions Rankings: The Most Comprehensive Ranking Approach to the World of Research Institutions

Benjamín Vargas Quesada[1], Atilio Bustos-González[2], and Félix de Moya Anegón[3]

[1]*Faculty of Communication and Documentation, Department of Information and Communication, Universidad de Granada, Spain*

[2]*SCImago Research Group, Chile*

[3]*SCImago Research Group, Spain*

Contents

9.1 Introduction

A picture is worth a thousand words. An image of science is suggestive, inspiring, and distinctive; it positions, places, informs at a glance, and invites to explore and encourage to know more. It is worth a million words. For this reason, the authors' research group has developed the Scimago Institutions Rankings (SIR), which is available at http://www.scimagoir.com/. It is a science visual evaluation resource to assess worldwide universities and research-focused institutions, because only rankings can offer much more information of quality than mere raw data.

The academic and research-related institutions are complex and must be treated as such. They cannot be evaluated by plebiscitary or survey-based systems. They need a complex indicator, understanding complexity in the sense of sophistication and compound, which takes into account all their features, effects, and impacts to be evaluated and ranked. Accordingly, the SIR Ranking is based on a composite indicator. It might seem trivial, because other indicators are too, but the SIR is much more than that. On the one hand, it is weighted by experts who decide which variables should be taken into account and how they are weighted and, on the other hand, by the emphasis placed by the experts as a result of the weighting. That is, what aspects have to be taken into account and which scientific aspects will weigh more. SIR considers scientific output as a proxy of academic quality. It is weighted more than the performance and technological impact (TI), because it judges scientific production before performance.

The final result is a classification ranked by a composite indicator that combines three different sets of indicators based on research performance, innovation outputs, and societal impact measured by their web visibility, which are all grouped in two dimensions: output and performance (see Figure 9.1). But neither of the two is sufficient, because the output can be higher or lower regardless of its performance, so the concept of performance should always be linked to scientific leadership (Moya-Anegón et al., 2013a,b,c, 2015).

It provides a friendly interface that allows the visualization of any customized ranking from the combination of these three sets of indicators. Additionally, it is possible to compare the trends for individual indicators of up to six institutions. For each large sector, it is also possible to obtain distribution charts of the different indicators.

SIR is now a *league table*. The aim of SIR is to provide a useful metric tool for institutions, policy makers, and research managers for the analysis, evaluation, and improvement of their activities, outputs, and outcomes.

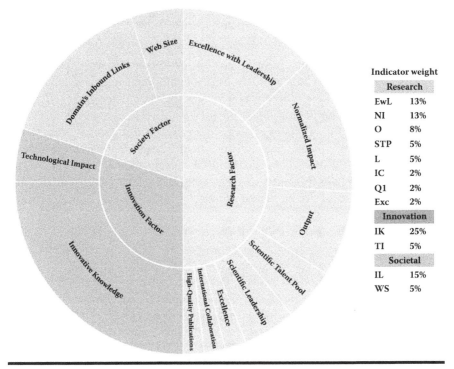

Indicator weight	
Research	
EwL	13%
NI	13%
O	8%
STP	5%
L	5%
IC	2%
Q1	2%
Exc	2%
Innovation	
IK	25%
TI	5%
Societal	
IL	15%
WS	5%

Figure 9.1 **Score indicators.**

9.2 General Considerations

For comparative purposes, the value of the composite indicator has been set on a scale of 0–100. However, the line graphs and bar graphs always represent ranks (lower is better, so the highest values are the worst).

9.2.1 Scimago Standardization

In order to achieve the highest level of precision for the different indicators, an extensive manual process of disambiguation of the institution's names has been carried out. The development of an assessment tool for Bibliometric analysis aimed to characterize research institutions involves an enormous data processing task related to the identification and disambiguation of institutions through the institutional affiliation of documents included in Scopus. The objective of Scimago, in this respect, is twofold:

1. Definition and unique identification of institutions: The drawing up of a list of research institutions where every institution is correctly identified and defined. Typical issues on this task include institution's merge or segregation and denomination changes.
2. Attribution of publications and citations to each institution: We have taken into account the institutional affiliation of each author in the field *affiliation* of the database. We have developed a mixed system (automatic and manual) for the assignment of affiliations to one or more institutions, as applicable, as well as an identification of multiple documents with the same DOI and/or title.

Thoroughness in the identification of institutional affiliations is one of the key values of the guaranteed standardization process which, in any case, achieves the highest possible levels of disambiguation.

Institutions can be grouped by the countries to which they belong. Multinational institutions, which cannot be attributed to any country, have also been included.

In the SIR website, the institutions marked with an asterisk consist of a group of subinstitutions, identified by the abbreviated name of the parent institution. The parent institutions show the results of all of their subinstitutions.

Institutions can be also grouped by sectors (government, health, higher education, private, and others). This filter enables analysis between similar institutions and may add other geographical constraints by regions or countries, for specific years in the range 2009–2016.

For ranking purposes, the calculation is generated each year from the results obtained over a period of 5 years and ending 2 years before the edition of the ranking. For instance, if the selected year of publication is 2016, the results used are those from the 5-year period 2010–2014. The only exception is the case of web indicators which have only been calculated for the last year. The window to select a ranking goes from 2009 to 2016.

The inclusion criterion is that the institutions had published at least 100 works included in the Scopus database during the last year of the selected period.

The source of information used for the indicators for innovation is the PATSTAT database.

The sources of information used for the indicators for web visibility are Google and Ahrefs.

9.3 Indicators

Indicators are divided into three groups intended to reflect scientific, economic, and social characteristics of research-related institutions. The SIR includes both size-dependent and size-independent indicators, that is, indicators influenced and not influenced by the size of the institutions, respectively. In this way, the SIR provides overall statistics of the scientific publication and other output of institutions, at the same time that enables comparisons between institutions of different sizes. It needs to be kept in mind that the final indicator has been calculated out of the combination of the different indicators (to which a different weight has been assigned) in which the resulting values have been normalized on a scale of 0–100.

9.3.1 Score Indicators

9.3.1.1 Research

The following are indicators under research:

1. *Excellence with leadership* (EwL): EwL indicates the amount of documents in excellence in which the institution is the main contributor (Moya-Anegón et al., 2013a,b) (size-dependent indicator).
2. *Normalized impact* (NI) (leadership output): NI is computed over the institution's leadership output using the methodology established by the Karolinska Institutet in Sweden named item-oriented field-normalized citation score average. The normalization of the citation values is done on an individual article level. The values (in decimal numbers) show the relationship between an institution's average scientific impact and the world average set to a score of 1; that is, an NI score of 0.8 means that the institution is cited 20% below world average and 1.3 means that the institution is cited 30% above average (González-Pereira et al., 2010; Guerrero-Bote and Moya-Anegón, 2012; Rehn and Kronman, 2008) (size-independent indicator).
3. *Output* (O): O is the total number of documents published in scholarly journals indexed in Scopus (Moya-Anegón et al., 2007; Lopez-Illescas et al., 2011; Romo-Fernández et al., 2011; OECD & Scimago Research Group (CSIC), 2016) (size-dependent indicator).
4. *Scientific talent pool* (STP): STP is the total number of different authors from an institution in the total publication output of that institution during a particular period (size-dependent indicator).
5. *Scientific leadership* (L): L indicates the amount of an institution's output as main contributor, that is, the amount of papers in which the corresponding author belongs to the institution (Moya-Anegón, 2012; Moya-Anegón et al., 2013a,b,c) (size-dependent indicator).
6. *International collaboration* (IC): IC is the institution's output produced in collaboration with foreign institutions. The values are computed by analyzing an institution's output whose affiliations include more than one country address (Guerrero-Bote et al., 2013; Lancho-Barrantes et al., 2012; Lancho-Barrantes et al., 2013; Chinchilla-Rodríguez et al., 2010, 2012) (size-dependent indicator).

7. *High-quality publications* (Q_1): Q_1 is the number of publications that an institution publishes in the most influential scholarly journals of the world. These are those ranked in the first quartile (25%) in their categories as ordered by SCImago Journal Rank (SJRII) indicator (Miguel et al., 2011; Chinchilla-Rodríguez et al., 2016) (size-dependent indicator).

8. *Excellence* (Exc): Excellence indicates the amount of an institution's scientific output that is included in the top 10% of the most cited papers in their respective scientific fields. It is a measure of high-quality output of research institutions (Bornmann et al., 2012; Bornmann & Moya-Anegón, 2014a; Bornmann et al., 2014b; Scimago Lab, 2016) (size-dependent indicator).

9.3.1.2 Innovation

The following are indicators under innovation:

1. *Innovative knowledge* (IK): IK is the scientific publication output from an institution cited in patents. It is based on PATSTAT (http://www.epo.org) (Moya-Anegón & Chinchilla-Rodríguez, 2015) (size-dependent indicator).

2. *TI*: TI is the percentage of the scientific publication output cited in patents. This percentage is calculated considering the total output in the areas cited in patents, which are the following: agricultural and/or biological sciences, biochemistry, genetics and molecular biology, chemical engineering, chemistry, computer science, earth and planetary sciences, energy, engineering, environmental science, health professions, immunology and microbiology, materials science, mathematics, medicine, multidisciplinary, neuroscience, nursing, pharmacology, toxicology and pharmaceutics, physics and astronomy, social sciences, and veterinary. It is based on PATSTAT (http://www.epo.org) (Moya-Anegón & Chinchilla-Rodríguez, 2015) (size-independent indicator).

9.3.1.3 Societal Impact

The following are indicators under societal impact:

1. *Domain's inbound link* (IL): IL is the number of incoming links to an institution's domain according to Ahrefs (https://www.ahrefs.com). This indicator is size dependent.

2. *Web size* (WS): WS is the number of pages associated to the institution's URL according to Google (https://www.google.com) (Aguillo et al., 2010). This indicator is size dependent.

Each of the indicators receives a differential percentage weight, which is shown in Figure 9.1. The weight of the components was determined by observation of the conduct of one thousand users, made between May 2014 and June 2016. In this period, the variable weight system was maintained open. The system learned weights were preferred by most. Shaped to this observation, the mixture finally adopted was determined.

9.4 Case Studies

The SIR composite indicator in combination with visualization techniques shows the image of science. As we have said earlier, a picture is worth a thousand words and the SIR a million. The proof is in (eating) the pudding.

A box-and-whisker plot diagram, also known as box plot, is a chart that provides information on the minimum and maximum values of a distribution, Q_1, Q_2 or median, Q_3 quartiles, and the interquartile range (IRC) or box (Q_1–Q_3). The whiskers, that is, the lines that go up and down from the box, extend to the maximum and minimum values of the series or up to 1.5 times the IRC. Values that extend beyond this IRC are called outliers.

Figure 9.2 shows the global picture of all higher education institutions, grouped by regions ordered by the composite indicator SIR, for the year 2016. The information from which this image has been constructed is available and can be downloaded from http://www.scimagoir.com/rankings.php?sector=Higher%20 educ.&country=all.

For better understanding of the figure, some of its elements are detailed and then are explained.

The X-axis of the image represents and groups the higher education institutions of each of the eight regions of the world collected by the SIR. The Y-axis corresponds to the position that each higher education institution has in the ranking, according to the composite indicator of SIR. In this case and for its better compression, the scale of this axis is shown inverted, that is, downward.

The minimum value of the distribution is represented by the higher education institution with the highest position of each of the regions of the world, in this case starting from above. It usually coincides with the first horizontal line of each case and with the beginning of the upper whisker. This is the case in all regions of the world, except in Africa, where the outliers, or atypical values of distribution, as we shall see, are above it. The value of Q_1 coincides with the second horizontal line of each region, just where the upper whisker ends and the dark gray color of the box begins. Q_2 or the median corresponds to the third horizontal line of the box, dividing the darkest and lightest gray areas, representing the central value of the distribution and indicating where 50% of higher education institutions are concentrated for each case. If the median is not in the center of the box, the distribution is not symmetric. Q_3 is represented by the horizontal line where the light gray color of each box ends and just where the lower whisker of the box begins. Finally, the maximum value usually coincides with the last horizontal line, just where the lower whisker ends, except in the case of Western Europe, where there are outliers.

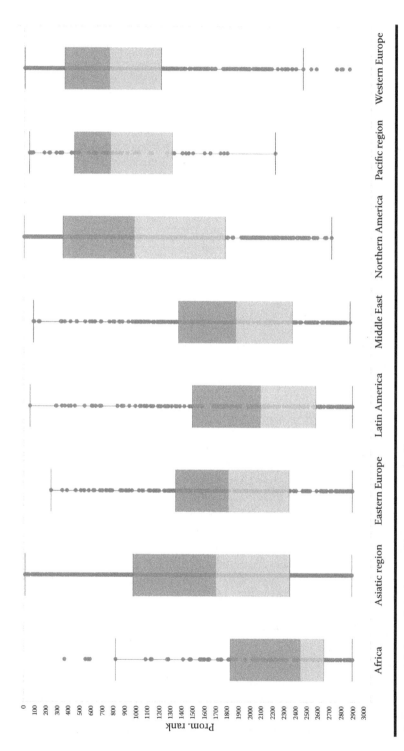

Figure 9.2   Higher education ranking per score indicator.

Having said that, we are able to comment on the figure of the higher education institutions ordered by their position, according to the composite indicator of SIR.

The first thing that catches the eye are the outliers that have the two regions at the ends of the figure. Africa has higher education institutions that are 1.5 times above its IRC, occupying relatively high positions in the ranking (within the best 500), although unfortunately, they are not the common thing, but the opposite. Paradoxically, in the region of Western Europe, there is the opposite effect, where we find higher education institutions with atypical positions in the lower area of the ranking or outliers, with values 1.5 times above their IRC.

Africa has a clearly asymmetric distribution. Its median (Q_2) is clearly below the middle of the box, occupying a very low position in the raking order. It is characterized by a reduced appearance and great dispersion of its higher education institutions in the high-ranking positions, as shown by its upper whisker, and a strong concentration in the lower positions of the lower whisker.

The Asian region offers a balanced distribution, with a large number of higher education institutions occupying relevant positions as seen in the upper whisker, but with a concentration of 50% of their higher education institutions (median), quite low when compared with other regions.

The Eastern Europe, Latin America, and Middle East regions are a mixture of the two previously seen. All three have very symmetrical distributions and with higher education institutions occupying, in some cases, high positions in the ranking, although with a great dispersion in their upper whiskers in all cases. The positions of its medians are low, even more than the Asian region, Latin America being the lowest of the three.

Northern America has a positive asymmetric distribution. That is, at least half of the positions of its higher education institutions are in the upper area of the ranking, as indicated by their median. Nevertheless, dispersion is observed in the lower whisker in its higher education institutions, indicating a blend of positions in the ranking.

The Pacific region also shows a positive asymmetric distribution. With a concentration of their higher education institutions (box) and median, in positions even better than those of Northern America, its problem is its strong dispersion throughout the distribution, not only in the whiskers as in other regions, but also in the box itself. In addition, it has some higher education institutions with quite low-ranking positions, which make its lower whisker quite long, which makes its maximum value descend too much.

Western Europe is very similar to the Northern American region. It has a positive asymmetrical distribution; its higher education institutions are in the top of the ranking, as well as its median, and very concentrated in a small box. Its problem is also in the dispersion of the positions of the lower whisker, where outliers or atypical values appear, indicating that few of its higher education institutions occupy the lowest positions in the ranking.

In general, the median regions in the highest-ranking positions and with a smaller box or higher concentration of its higher education institutions are Western Europe, the Pacific region, and Northern America, in this order.

As you have seen, the informative quality of this image is very good. However, the balance obtained in the ranking of higher education institution positions results in few differences between regions, although there really are large differences. Although these values, such as whiskers, are quite informative by themselves, this figure would need to be complemented with another to get an even more complete, revealing, and informative view of higher education institutions in a more individualized way. It is necessary to obtain a figure with the representation of values of SIR Ranking, that is to say, the distances that exist among higher education institutions. If not, a visual effect can be produced that leads us to deduce that the higher education institutions of one region and others are alike or very similar, when actually, the differences can be abysmal.

Figure 9.3 is obtained by representing the value that each higher education institution gets in the composite indicator of SIR. In our opinion, this is the one that best reflects reality.

The aspects to consider for the analysis and interpretation of this figure are the same as for the previous one. The difference is that in this case, the *Y*-axis represents the values obtained by the higher education institution in SIR, not its position, and that the scale of the axis is not inverted; that is, it is shown in an upward direction.

Proportionally, in terms of the position of medians, or where 50% of higher education institutions concentrate, and the size of boxes or IRC, the second figure offers information very similar to the first one, although on a different scale. The first figure helped us to visualize the global context of science through its higher education institutions, while the second one provides us with comprehensive, detailed, and contextual information about each of them. Therefore, both figures are necessary, since its combination offers a more complete view of science.

But what is really important about the second figure is the information that it gives about the large number of outliers in the upper whisker in each and every region of the world (in the first figure, they were identified very lightly only in the case of Africa). That is, it highlights and identifies the higher education institution with positive atypical values or, what is the same, those that are 1.5 times above its IRC, which is the true responsibility of giving prestige to each region, because it is the most relevant worldwide and makes the difference with the rest.

For example, the first higher education institution of SIR Ranking, with a value of 40.50, is Harvard University, from the Northern American region. In addition, the second, although very far from this and with a value of 22.51, is Stanford University, also from the region of Northern America. The third is Massachusetts Institute of Technology with 18.77; the fourth, the University of Michigan, Ann Arbor, with 18.77; and the fifth, Johns Hopkins University, with 18.21, which are also from the Northern American region. It is necessary to descend to the sixth position of the ranking to find the first of the Western Europe region: the University of

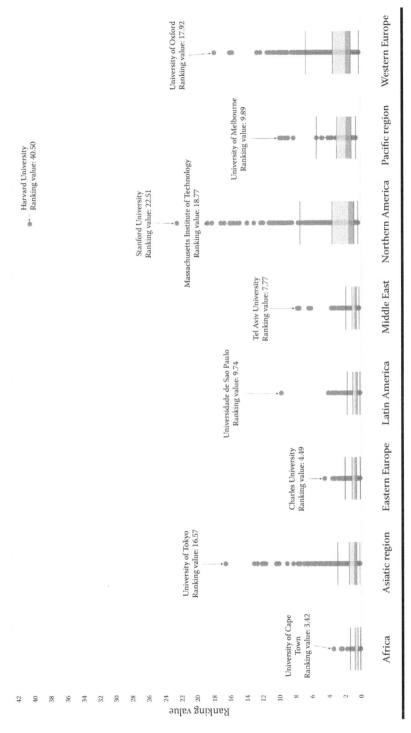

Figure 9.3 **Higher education ranking per value indicator.**

Oxford with a value of 17.92, which is, in turn by far, the one that obtains a better position in this region.

Therefore, through this image of Figure 9.3, it is possible to identify at a glance the higher education institutions that stand out worldwide and in each region, that is, those that are proxies of academic quality.

All this is possible only because of the emphasis of experts. As we have said before, the composite indicator of SIR emphasizes high-quality scientific production and cutting-edge research. For this reason, the higher education institutions that are intensive in research benefit in the ranking, since in it, the high value of scientific production is considered to be a proxy of academic quality, precisely because it has that emphasis. For all of this, we say and maintain that *rankings are, far more than just raw data.*

References

Aguillo, I., Bar-Ilan, J., Levene, M., Ortega, J. (2010). Comparing university rankings. *Scientometrics*, 85 (1), pp. 243–256.

Bornmann, L., de Moya Anegón, F., Leydesdorff, L. (2012). The new excellence indicator in the World Report of the SCImago Institutions Rankings 2011. *Journal of Informetrics*, 6 (2), pp. 333–335.

Bornmann, L., & de Moya Anegón, F. (2014a). What proportion of excellent papers makes an institution one of the best worldwide? Specifying thresholds for the interpretation of the results of the SCImago Institutions Ranking and the Leiden Ranking. *Journal of the Association for Information Science and Technology*, 65 (4), pp. 732–736.

Bornmann, L., Stefaner, M., de Moya Anegón, F., Mutz, R. (2014b). Ranking and mapping of universities and research-focused institutions worldwide based on highly-cited papers: A visualization of results from multi-level models. *Online Information Review*, 38 (1), pp. 43–58.

Chinchilla-Rodríguez, Z., Vargas-Quesada, B., Hassan-Montero, Y., González-Molina, A., Moya-Anegón, F. (2010). New approach to the visualization of international scientific collaboration. *Information Visualization*, 9 (4), pp. 277–287.

Chinchilla-Rodríguez, Z., Benavent-Pérez, M., Miguel, S., Moya-Anegón, F. (2012). International collaboration in medical research in Latin America and the Caribbean (2003–2007). *Journal of the American Society for Information Science and Technology*, 63 (11), pp. 2223–2238.

Chinchilla-Rodríguez, Z., Zacca-González, G., Vargas-Quesada, B., Moya-Anegón, F. (2016). Benchmarking scientific performance by decomposing leadership of Cuban and Latin American institutions in public health. *Scientometrics*, 106 (3), pp. 1239–1264.

González-Pereira, B., Guerrero-Bote, V., Moya-Anegón, F. (2010). A new approach to the metric of journal's scientific prestige: The SJR indicator. *Journal of Informetrics*, 4 (3), pp. 379–391.

Guerrero-Bote, V.P., & Moya-Anegón, F. (2012). A further step forward in measuring journals' scientific prestige: The SJR2 indicator. *Journal of Informetrics*, 6 (4), pp. 674–688.

Guerrero-Bote, V.P., Olmeda-Gomez, C., De Moya-Anegón, F. (2013). Quantifying the benefits of international scientific collaboration. *Journal of the American Society for Information Science and Technology*, 64 (2), pp. 392–404.

Lancho-Barrantes, B.S., Guerrero-Bote, V.P., Chinchilla-Rodríguez, Z., Moya-Anegón, F. (2012). Citation flows in the zones of influence of scientific collaborations. *Journal of the American Society for Information Science and Technology*, 63 (3), pp. 481–489.

Lancho-Barrantes, B.S., Guerrero-Bote, V.P., de Moya-Anegón, F. (2013). Citation increments between collaborating countries. *Scientometrics*, 94 (3), pp. 817–831.

Lopez-Illescas, C., de Moya-Anegón, F., Moed, H.F. (2011). A ranking of universities should account for differences in their disciplinary specialization. *Scientometrics*, 88 (2), pp. 563–574.

Miguel, S., Chinchilla-Rodríguez, Z., Moya-Anegón, F. (2011). Open Access and Scopus: A new approach to scientific. From the standpoint of access. *Journal of the American Society for Information Science and Technology*, 62 (6), pp. 1130–1145.

Moya-Anegón, F., Chinchilla-Rodríguez, Z., Vargas-Quesada, B., Corera-Álvarez, E., González-Molina, A., Muñoz-Fernández, F.J., Herrero-Solana, V. (2007). Coverage analysis of Scopus: A journal metric approach. *Scientometrics*, 73 (1), pp. 57–58.

Moya-Anegón, F. (2012). Liderazgo y excelencia de la ciencia española. *Profesional de la Información*, 21 (2), pp. 125–128.

Moya-Anegón, F. (dir.), Chinchilla-Rodríguez, Z. (coord.), Corera-Álvarez, E., González-Molina, A., Vargas-Quesada, B. (2013a). Principales Indicadores Bibliométricos de la Actividad Científica Española: 2010. Madrid: Fundación Española para la Ciencia y la Tecnología. http://icono.fecyt.es/informesypublicaciones/Documents/indicadores%20bibliometricos_web.pdf

Moya-Anegón, F. (dir.), Chinchilla-Rodríguez, Z. (coord.), Corera-Álvarez, E., González-Molina, A., Vargas-Quesada, B. (2013b). Excelencia y liderazgo de la producción científica española 2003–2010. Madrid: Fundación Española para la Ciencia y la Tecnología. http://www.scimagoir.com/pdf/iber_new/SCImago%20Institutions%20Rankings%20IBER%20es.pdf

Moya-Anegón, F., Guerrero-Bote, V.P., Bornmann, L., Moed, H.F. (2013c). The research guarantors of scientific papers and the output counting: A promising new approach. *Scientometrics*, 97 (2), pp. 421–434.

Moya-Anegón, F. (dir.), Bustos-González, A. (coord.), Chinchilla-Rodríguez, Z., Corera-Álvarez, E., López-Illescas, C., Vargas-Quesada, B. (2015). Principales indicadores cienciométricos de la actividad científica chilena 2013. Informe 2015. Santiago: CONICYT. http://www.informacioncientifica.cl/Informe_2015/

Moya-Anegón, F., Chinchilla-Rodríguez, Z. (2015). Impacto tecnológico de la producción universitaria iberoamericana. En: La transferencia de la I+D, la innovación y el emprendimiento en las universidades. Educación Superior en Iberoamérica. Informe 2015. Santiago de Chile: Centro Interuniversitario de Desarrollo, pp. 83–94. https://www.redemprendia.org/sites/default/files/descargas/informeTransferencial%2BD2015.pdf

OECD and Scimago Research Group (CSIC-Spain). (2016). *Compendium of Bibliometric Science Indicators*. OECD, Paris. Accessed from http://oe.cd/Scientometrics

Rehn C, Kronman U. (2014). *Bibliometric Handbook for Karolinska Institutet*. Karolinska Institutet University Library. Version 2.0. Solna, Sweden: Karolinska Institutet. https://kib.ki.se/sites/default/files/bibliometric_handbook_2014.pdf

Romo-Fernández, L.M., Lopez-Pujalte, C., Guerrero Bote, V.P., Moya-Anegón, F. (2011). Analysis of Europe's scientific production on renewable energies. *Renewable Energy*, 36 (9), pp. 2529–2537.

Scimago Lab. (2016). Scimago Institution Rankings (SIR). http://www.scimagoir.com

Chapter 10

Knowledge Distribution through the Web: The Webometrics Ranking

Bárbara S. Lancho-Barrantes

Tecnológico de Monterrey, Scientometrics Research Monterrey, Mexico

Contents

10.1 Introduction

In order to fully understand Webometrics, it is important to remember certain data, without having to go into excessive detail of the main engine that makes possible its existence, the web. In the same way it is necessary the framing of this discipline within those (Informetrics, Bibliometrics, Scientometrics, etc.) of which it has nurtured to reach the degree of autonomy that contemporarily possesses. With a view to comprehend the functioning and meaning of Webometrics, it is essential to explain the mechanisms that exist today to carry out the posed types of analysis. It is equally obligatory to name the main Webometrics ranks that exist so far, since they are currently fundamental resources and sources of consultation and study for researchers and society in general. Finally, in order to show a general current perspective of the scientific production in Webometrics that has been carried out and is being performed all over the world, a sample extracted from the international bibliographic database *Scopus* is presented, with a publication window from 1997 to 2016, in such a way that by means of figures and tables that will be shown, among other results, which scientists, which institutions, which countries, and which disciplines publish more works related to the present matter.

10.2 World Wide Web

The web is the main means that makes available to the users the wide network of networks and sources of global information that is the Internet. It was born and was developed by the Englishman Tim Berners-Lee from 1989 at the Center Européenne pour la Recherche Nucléaire (CERN) in Geneva, Switzerland. The meaning of World Wide Web (whose simplification web is *WWW* or even *W3*) has been translated as *global web*. The official propaganda put forward by CERN defines it as *distributed hypermedia system* (Boutell, 1994). It has its own communication protocol called *hypertext transfer protocol* (HTTP), which determines the way in which WWW documents are transmitted through the network.

Following the definition gathered in the online Webopedia, WWW is conceived as a system of Internet servers that mainly supports formatted documents.

The documents are written in a language called *hypertext markup language* (HTML), which allows links to other documents, such as graphics, audios, and video files.

The W3 Consortium, understood as the international consortium that generates recommendations and standards that ensure the long-term growth of WWW, defines the web as a collection of intertwined pages, including a home page, and housed in the same locating network.

Throughout the distributed hypermedia architecture of the web, it works as an *associative memory* with capacity for learning by the strengthening of the links that are used much more frequently (Heylighen & Bollen, 1996; Faba-Pérez et al., 2004).

Regarding the characteristics of these links, Baron et al. (1996) were the first to divide the links into two types within hypertextual documents: organizational links, which expose the sequence in the structure in the same way than the structural links, and content-based links that are subdivided into semantic, rhetorical, and pragmatic. Focusing on a more technical field, it is necessary to highlight the work of García-Santiago (2001) that categorizes the links according to inherent links, which are those that refer to other files in order to deploy a single web page; they are identified by the HTML tag; internal links, which are responsible, via URL, to connect with other pages or files that are located in the same web; and external links, whose objective is to send pages or files located in another web and with a different URL (Orduña-Malea & Ontalba-Ruipérez, 2013).

10.3 Historical Background of Webometrics

This large miscellany of information accessible on the web allows us to consider it as a huge source of substantial, heterogeneous, unstructured, and distributed information. Since the web was conceived as a fundamental information resource capable of being analyzed and quantified, Webometrics arises as an emerging need.

The link between the growing and accelerated development of electronic information, the broad potential of fledgling technologies, and the burgeoning media has drawn the attention of researchers, who have conducted a number of studies to measure, in a quantitative point, information enclosed in the WWW.

In order to understand the etymology of the term *Webometrics*, a brief summary of the disciplines, which has needed to be nurtured to comprise itself as an independent domain, will be shown. In Figure 10.1, Webometrics is contextualized within the different metric disciplines.

The foundation of the Informetrics field (German: *Informetrie*) is attributed to Otto Nacke, who proposed it for the first time in 1979. Informetrics includes the analysis of quantitative aspects of information, regardless of how it appears registered and how it is generated. It also takes into account the quantitative aspects of informal communication and analyzes in depth the needs and uses of information for any activity, whether intellectual or not. On the other hand, it can incorporate and use various means in the measuring of the information.

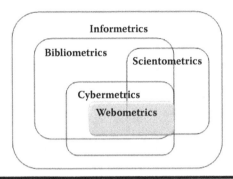

Figure 10.1   Interdependence between Informetrics–Bibliometrics–Scientometrics–Cybermetrics–Webometrics within the information science area. Geared by the model of Björneborn (2004).

In 1969, Alan Pritchard proposed a definition for Bibliometrics, in an article in the *Journal of Documentation* entitled "Statistical Bibliography or Bibliometrics?" that it is to be understood as the application of statistical and mathematical methods to define the processes of written statement, the origin and development of the scientific disciplines by means of counting techniques, and analysis of the different aspects of this communication.

Bibliometric studies aim at the quantitative assessment of science through scientific publications, which is why they currently provide the fundamental tool for the evaluation and knowledge of the research science activity, thus contributing to the scientific situation of a country, an institution, or a scientific discipline, which makes it possible to evaluate the efficiency of scientific activity and its impact on the community. From Bibliometrics emerges the Bibliometric indicators, based on scientific publications that are born with the purpose of adding greater numerical objectivity to the process of evaluation of the activities implemented in the field of research and science, providing further information on the results of the research process, valuing scientific activity, and the influence of both work and sources. The essential objective of this type of measures has been, since its inception, to analyze the production and scientific results. It is a usual standard practice to apply these types of indicators for evaluation purposes. The use of Bibliometric indicators presents a wide collection of assets over other methods that are used in scientific evaluation, since it is an objective, neutral, and verifiable method, the results of which are reproducible. In addition, these indicators can be used to analyze a large volume of data (Lascurain, 2006; Velasco-Gatón et al., 2012).

Parallel to the consolidation in Western countries of the term *Bibliometrics* in the early 1970s, in the countries of the former Soviet Union, the term *Scientometrics* stems from Russian *naukometry*, which in turn was derived from *naukovodemia* (Russian name for science), with certain overlapping with Bibliometrics. The linkage between the rise of matter, measurement, and the study of science has been clearly manifested in the birth and trajectory of *Scientometrics* journal, founded in Hungary in

1978. This journal is located in the first quartile of impact of the subject area library and information science and is edited by Wolfgang Glänzel.

Scientometrics can be considered as one of the most important within the metrics since it studies the quantitative aspects of science as a discipline or economic activity, is part of the sociology of science, and finds application in the establishment of scientific policies, where it includes, among others, the publication. Scientometrics, such as the other two disciplines studied, employs metric techniques for the evaluation of science (the term *science* refers to both the pure and social sciences) and evaluates the development of the scientific policies of countries and organizations.

Cybermetrics was used for the first time by Ali Ashgar Shiri in 1998 at the 49th Federation for Information and Documentation Congress and Conference held in New Delhi. Cybermetrics provides a set of knowledge and techniques based on Bibliometric and computational approaches, thus allowing the collection and analysis of cyberspace (or network space) in order to obtain all the information related to the contents, resources, and users in the network for this purpose to know and predict patterns of use and consumption of information (Orduña-Malea & Aguillo, 2014).

This discipline would embrace statistical studies of focus groups, mailing lists, or e-mail, including, of course, the web. Other terms related to *Cybermetrics* that have emerged over the years include *Netometrics* (Bossy, 1995), *Internetometrics* (Almind & Ingwersen, 1996), metrics of influence or *Influmetrics* (Cronin, 2001), and *Netinformetrics* (Orduña-Malea, 2012). There is a scientific journal called *Cybermetrics*, which is in the first quartile of impact of the area of library and information science. Since 1997, it has been publishing works dedicated to this subject in electronic format and in English language.

Finally, I will focus on the definition of Webometrics. This term was used for the first time in a study by Almind & Ingwersen (1997) about the presence of Danish universities on the web, so they attempted to introduce the application of Informetric and Bibliometric methods to WWW under the name *Webometrics*. According to Björneborn (2004), Webometrics is defined as the study of quantitative aspects of the construction and use of information resources, structures, and technologies on the web, based on Bibliometric and Informetric approaches, and collection of the construction and use of the main traits: (a) analysis of contents of the web pages, (b) analysis of the structure of the web links, (c) analysis of the use of the web (including the study of log files of research and browsing of users), and (d) web technology analysis, including search tool results.

The definition clearly reflects the great correlation with Bibliometrics; other similar terms that have emerged are *Webometry* (Abraham, 1996), *Web Bibliometry* (Chakrabarti et al., 2002), and *Web Metrics* (the term used in computer science by Dhyani et al., 2002). Notwithstanding the precise delimitation between Webometrics and Cybermetrics, at the moment, they are often used as synonyms. In fact, the term *Cybermetrics* is used to refer purely to Webometric studies. At the same time, and from the field of computer science, terms such as *cybergeography* or *cybercartography, web ecology, web mining,* web *graph analysis,* and *web intelligence* have arisen in order to designate certain modalities related to this kind of studies.

10.4 Practical Application of Webometrics

Thelwall et al. (2005) simplified the definition by stating that Webometrics is "the quantitative study of Web-related phenomena," embracing the research that goes further more than the information science. Even though Webometrics draws on Bibliometric sources, it must be able to develop methods, tools, and techniques that allow the identification, observation, and quantification of social changes that are demonstrated in the web.

The analysis of citation on the web is based on the equitable comparison between citation and links. Reference link is spoken to refer to those links made from the website under study and citations for the received links (McKiernan, 1996). Mike Thelwall (2003), in an article entitled "What Is This Link Doing Here? Beginning a Fine-Grained Process of Identifying Reasons for Academic Hyperlink Creation," tried to find the motivations that are exclusively of the links concerning the traditional citations. His conclusions are similar to those already proposed by Kim (2000), who assured that electronic links respond more to a scientific-social-technological behavior than to an exclusive striving of scientific persuasion.

10.5 Tools for Extracting Data from the Web

The essential thing to perform a web-based analysis is to extract the information of the web that is going to be examined, that is to say, to extract the data that are needed to be able to apply indicators and to be able to examine the obtained results. This section will show the different methods that exist for the extraction of data. The more complete is the information obtained, the web-based study will be more comprehensive and will have better quality.

The extraction of quantitative data has become a great difficulty, and this is due to the huge volume of information that today cohabits on the web. Therefore, it is imperative to use automated methods, both in the collection and in the analysis of the results, in order to directly obtain information from a number of previously chosen pages.

The data collection is executed thanks to crawlers, web spiders, or web robots, which are specialized software programs that are in charge of running the web in a methodical, systematic, and automated way, in general exploring the hypertextual tree by compiling, overturning the contents, and indexing all the text that they find, thus forming very broad databases in which the Internet users will later carry out their searches by including the keywords. Crawlers, such as any other types of software, are used for various purposes, although the most recurring use is the software agent search engines. The most popular crawlers are the ones that provide the main search engines (such as Google, Yahoo, Bing, Exalead). This procedure of using a crawler in the main search engines called indirect data extraction methods is in fact the most powerful and complete with a difference. Although for a large majority of users, crawlers are conceived as a synonym for search engines, the possibilities that a crawler can offer go much further.

Nowadays the possibility of creating specific crawlers rises significantly. Many companies program specific crawlers for data extraction as a business. Therefore, this opens a new market for web developers in which the design of crawlers with particular high relevance tasks become a new work field. Hence, this direct mode of extracting the data is called direct extraction methods, and they are much simpler. Some well-known web crawlers are SocSciBot, Astra Site Manager, Funnel Web Profiler, Xenu Link Sleuth, and Linkbot.

10.6 Methods of Direct Data Extraction

SocSciBot is the best web crawler to explain this method, since it is a web crawler intended for the investigation of link analysis in a website or collection of websites or for the search/analysis of text in a collection of sites designed by the Statistical Cybermetrics Research Group of the University of Wolverhampton in the United Kingdom. SocSciBot is composed of a tracking module, a link analysis module, and a text analysis module. The analysis research carried out consists of two phases: a first tracking, which can be determined by a single website or several through a list of seeds, and the second phase, where it could be either study the links of the crawled sites or the contents of the pages retrieved through the analysis module of the cyclist content. In the link analysis module, it is mandatory to choose what kind of links that are to be included in the analysis: links within a website (site self-links), links between sites introduced as seeds (intersite links of crawled sites), or links between the sites belonging to the group of seeds and sites not belonging to the seed group (intersite links of sites not crawled).

In the first tab of the link analysis module, the main reports that SocSciBot offers are as follows: *Page and link counts*: the number of pages and outbound links in each site; *ADM count summary*: the number of incoming and outgoing links for each site using the four types of URL groupings (page, directory, domain, or site); *All external links*: a list of the external links of each site grouped by the four options already indicated and the outbound links directed to sites not belonging to the group of sites tracked; *Directory interlinking*: the number of links between directory pairs in different sites; *Domain interlinking*: the number of links between domain pairs in different sites; *File interlinking*: the number of links between pairs of web pages in different sites; *Site interlinking*: the number of links between pairs of sites; *Selected external links*: the lists of the outgoing links according to the type of grouping chosen in the initial phase of the link analysis, remember: page, directory, domain, or site; *Selected external links with counts*: the lists of the outgoing links and the frequency of each one, also depending on the level of grouping; *Unselected external links*: provides and shows a list of outgoing links to pages, directories, domain, or sites (depending on our choice) not belonging to the crawled site group; and *Unselected external links with counts*: supplies a list of outgoing links to pages, directories, domain, or sites (according to our choice) not belonging to the group of sites tracked with the frequencies of each of them.

In the same module also, the network of links between the sites already traced, the network of outbound links from the one-to-one sites, or the network of links including links to untracked sites can be viewed.

All of these options can be displayed through the SocSciBot display module or using Pajek, UCINET, or any other software that uses the .net format. The *visualization* module used by SocSciBot is the Webometric Analyst Network Graph Tool. In the content analysis module, also termed *cyclist*, it allows indexing the contents of the pages already recovered during the crawl phase to create a linguistic corpus. To do this, there will be a request to enter as input a text file with a list of *empty words* or *stop words*; these are words that have no semantic content such as prepositions, conjunctions, and articles. In addition to the possibility of eliminating empty words, cyclist applies a stemming or a basic lemmatization by eliminating plurals and words that contain numbers.

Once the content is indexed and the corpus is created, it is possible to search for words using the search tool that incorporates this module. The search tool will display the indexed pages containing the searched word by providing the results with the supposed URL, the HTML of the page, and the first 100 words of the page.

In this module, reports that contain the total number of words that comprise the corpus and how often each word appears or filter all words that contain a specific root or text can be created. It is another tool developed by the Statistical Cybermetrics Research Group of the University of Wolverthampton as Webometric Analyst. This tool, unlike the SocSciBot crawler, is based on the results obtained through the Bing Search API (Thelwall & Sud, 2012) to offer several types of Webometric analysis including platforms such as Twitter, YouTube, or Mendeley.

10.7 Methods of Indirect Data Extraction: Through Search Engines

Search engines or tools for searching are huge databases generated as a result of the automatic indexing of documents previously stored on the web. These search engines collect documents in HTML format and other types of resources, to carry out their operation using a robot. Search bots, also called spiders, wanderers, web ants, crawlers, or worms, roam the various servers in a recursive way, through the links provided by the pages, that they are descending it as if it was from a tree, so it was treated through the different branches of each server.

The best advantage of using a search engine is speed and immediacy, but it is a great disadvantage that it can be difficult to find the things being searched if the query language to the database is not mastered.

A search engine is composed of the following elements:

- A program that explores the network (robot) in order to locate documents and addresses of documents. So, it searches periodically on the web servers.
- A database that includes an automatic indexing system and an interrogation system with a query language.
- An interface: The browsers allow the search through an interface, in other words, one that allows the interrogation of the system and obtains the desired outcomes. The searches are performed through the introduction of a number of operators, including Boolean operators (AND, NOT, OR, XOR), proximity operators (NEAR, ADJ, +), absence operators (–), truncation, and parentheses.

Presently, there are advanced search operators such as Google, Exalead, Gigablast, Bing (ex-Live, ex-MSN Search), and Ask, which allow delimiters per domain, which makes it possible to use them for web-based purposes. An example of recovery in Bing is interlinking linkfromdomain: csic.es site: wikipedia.org.

In Figure 10.2, an example of Bing's advanced search with the operator *link from domain: csic.es site: wikipedia.org* is shown.

In Figure 10.3, an example of the total number of external links received by a site is displayed.

Figure 10.2   Bing's advanced search operator *linkfromdomain* finds all linked pages from a given domain.

Figure 10.3   Example of how to search pages that address a URL through Exalead: external links–visibility.

More often than not, reliable results can be obtained from the APIs, which offer different search engines although it must be borne in mind that they work on databases that are less up to date and, generally, smaller than the commercial interfaces.

10.8 Recovery Algorithm: PageRank

Google, a popular search engine, uses the PageRank system to help determine the importance or relevance of a web search.

The following is the PageRank formula of Brin and Page (1998):

$$PR\,(A) = (1 - d) + \left(PR(Ti)/C(Ti) + \ldots + PR\,(Tn) + C(Tn)\right).$$

where
  $PR(A)$ is the PageRank of page A,
  $PR(Ti)$ is the PageRank of pages Ti that link to page A,
  $C(Ti)$ is the number of outbound links on page Ti, and
  d is a damping factor that can be set between 0 and 1.

PageRank trusts in the democratic nature of the web by using its wide relay structure as an indicator of the value of a particular page. Google interprets a link from page A to page B as a vote, from page A to page B.

However, Google not only goes beyond the volume of votes, or the links that a page receives, but also analyzes the page that issues the vote. Pages that are considered *important*, that is, with a high PageRank, that cast certain votes are worth more and help make other pages important. Therefore, the PageRank of a page reflects the importance of the same on the Internet.

10.9 Webometrics Studies

A web-based analysis can have a wide range of objects: regional, national, international, and so forth. It can be done in thematic areas, focusing on certain institutions or taking into account a time span. As discussed earlier, the sources employed are the search tools, the programs ad hoc, the rates, the network, and so forth. The applied methodology is based on an extensive typology of resources. These studies can be applied classical Bibliometric laws, such as the law of Zipf, law of distribution of the frequencies of utilization of the words in the texts; Lotka's law, quadratic inverse law of the scientific productivity; and the Bradford's act, law of dispersion of the scientific literature.

10.10 Webometrics Indicators

These indicators can be divided in two parts: *descriptive indicators*, which involve the explicit information (analyses of the features), and *content indicators*, which involve the relations and the implicit information (link analyses).

10.10.1 Descriptive Indicators

Descriptive indicators are the techniques employed for the evaluation of the design and the special features of the web spaces. The models of features are *subjective features* and *formal features*. A tool already known for the analyses of the formal feature web is Linkbot.

Another indicator linked with formal features of the web is the indicator of web accessibility: Web Accessibility 2.0 is explained as the universal access to the web, aside from the type of hardware, software, network infrastructure, language, culture, geographic location, and user capability. A tool employed for ensuring the degree of the web accessibility is named TAW (Web Accessibility Test), which is a tool developed by the Centre for the Development of Information and Communication Technologies in Asturias, which allows to automatically ensure some aspects of the web accessibility.

An accessible web must be *transformable*: the information and the services must be accessible for all the users who need them and they must be used with all the navigation displays; *comprehensible*: they must have clear and simple contents; and *navigable*: they must have simplified tools of navigation.

10.10.2 Content Indicators: Analysis of Citations on the Web: Indicators for Link Analyses

In the mid-1990s, several articles exposed the recognition of structural similarities between a network of linked web pages and a network of scientific documents linked by citations (Bossy, 1995; Larson, 1996).

In the studies of Björneborn (2004) and Björneborn & Ingwersen (2004) could be found some words related with links comparing them with Bibliometrics terms: Inlink, Outlink, SelfLink, Co-links, Co-outlinks, or Internal Links and External Links. Inlinks (also known as incoming links, inbound links, inward links, back links) are those links that are received by a node within the network, and on the other hand, Outlinks are those that, unlike Inlink, point to other pages. Thus, making a similarity with Bibliometrics, the Inlinks are conceived as the citation and Outlinks are understood as the references. Therefore, both conceptions are related to the direction in which they are given by the node or by the unit of information. The SelfLink, in terms of Bibliometrics, is conceived as the self-citation or the Co-link or co-citation (linking two nodes or being linked by two nodes at the same time) and is related to the function that each link carries within nodes

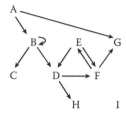

Figure 10.4 Topology of the links on the web. (From Björneborn, L., Small-World Link Structures across an Academic Web Space: A Library and Information Science Approach, Department of Information Studies, Royal School of Library and Information Science, Copenhagen, 2004.)

of the cluster. The term *Co-outlink* is equivalent in Bibliometrics to bibliographic coupling.

In 1996, McKiernan introduced the concept of citation, which was later adopted by Rousseau, specifically in 1997, who introduced a new element prior to the development of different Webometric indicators. That is why the analysis of links is considered as the cornerstone of Webometrics (Aguillo, 2012).

In Figure 10.4, the topology of the links on the web which was elaborated by Björneborn in 2004 can be seen.

- B receives a sitation (Inlink) from A.
- B has a link (Outlink) to C.
- B receives a SelfLink.
- A does not receive link.
- C has no outbound links.
- E and F are linked reciprocally.
- I is isolated.
- D, E, and F are linked transitively.
- A has a transverse output link to G (cutoff).
- H is reached from A by a path of direct links.
- C and D are colinked by B.
- B and E are linked to D (bibliographic coupling).

10.11 Web Impact Factor

In order to interpret the web impact factor (WIF) it is necessary to make a brief reference Impact Factor (IF). It is therefore conceived as the best-known and most valued indicator of quality by the evaluation bodies, which is given by the JCR tool. It is based on the number of times, on average, in which an article published in a particular journal is cited; that is, it is calculated by dividing the number of citations in the current year of articles published in the previous 2 years between the numbers, overall, of the published articles in these two previous years. This concept of IF was suggested by Garfield (1955).

The WIF concept, proposed by Ingwersen (1998), is defined as the number of links (internal and external links) that enter a web space, divided by the number of pages found in that web space at any given time. Different types of WIFs: total, external, and internal (Smith, 1999) can be found. The external IF is the indicator that best exposes the visibility of a website. The numerator of this formula is obtained by using the link command. The denominator is obtained by using the site command.

$$WIF = \frac{link:URLx}{site:URLx}.$$

A real example of external WIF calculation is as follows:

$$WIF\ external = \frac{link:csic.es}{site:csic.es}.$$

10.12 Webometrics Ranking

There are interesting web-based projects developed by the Cybermetrics Lab research team, which is part of the Spanish National Research Council (CSIC) and led by Isidro F. Aguillo, which are being carried out several years ago and are called the Webometrics Rankings. Several types can be found and will be explained hereunder.

The rankings are as follows: Ranking Web of Universities, Ranking Web of Repositories, Ranking Web of Hospitals, Ranking Web of Research Centers, Ranking Web of Business School, and Ranking Web of Researchers.

The main objective of this kind of rankings is to promote web publishing and support open access initiatives, electronic access to scientific publications, and other academic materials as main objectives. Nonetheless, web indicators are very useful for classification work, as they are based on the overall performance and visibility of universities. In addition to formal publications (electronic journals, repositories), the web includes informal academic communication. Publishing on the web is cheaper, and the high-quality standards of peer review processes are maintained.

It could also reach much wider potential audiences by providing access to scientific knowledge to researchers and institutions located in developing countries and to third parties (economic, industrial, political, or cultural applicants) on their own community. The Webometrics Rank grants a 50% weight to the volume of information published on the web and the other 50% to the links received by such pages. In other words, a 1:1 ratio is maintained between size and visibility. A second level takes into account the specific contents of a web, and the model reinforces in its weights the contribution of documentary formats (rich files) and specifically scientific articles and materials in relation to them. Consequently, a redistribution of weights of 50% corresponding to the activity is mandatory.

10.12.1 Ranking Web of Universities (http://www.webometrics.info/en)

Ranking Web of Universities has been published since 2004, is directly inspired by the model of the Shanghai Ranking, but uses cybermetric data that are extracted from the websites of the universities.

It also has a great international coverage, and the data of all the institutions can be consulted (in which there are more than 21,000 currently) that come from the search engines in the Internet and the scholar database of GS that are also in open access.

As with other international rankings, consultations by areas, geographic regions, and groups of countries of emerging economies and markets, such as BRICS and CIVETS (Colombia, Indonesia, Vietnam, Egypt, Turkey, and South Africa) are fully permitted.

The indicators used to rank universities are the following:

■ *Presence rank*: It is the total number of web pages, including all the rich files (such as PDF), according to Google. The weight assigned is 10%.
■ *Visibility/impact rank*: It is used as the maximum indicator value obtained from two independent Inlinks information providers (Ahrefs and Majestic). The weight assigned is 50%.
■ *Openness rank*: Now it is going to be named *transparency*, and it will use the data from the GSC institutional profiles. Only institutional profiles are chosen. The weight assigned is 10%.
■ *Excellence rank*: It is the Scimago data (top 10% most cited papers by discipline) for a period of 5 years 2010–2014. The new weight is 30%.

10.12.2 Ranking Web of Repositories (http://repositories.webometrics.info/)

The primordial objective of Ranking Web of Repositories is to promote open access initiatives. A relevant issue for the universities and the research centers is the deposit of scientific articles and related material in the institutional or thematic repositories. This Ranking Web of Repositories consists of a list of repositories that mainly host the research works, which are classified according to a composite indicator that combines data of web presence and web impact (hypertextual visibility) all coming from the major search engines.

The publication of the Ranking Web of Repositories was launched in 2008 and is updated every 6 months, approximately in late January and July. There are about 2,275 repositories collected in the world ranking.

10.12.3 Ranking Web of Hospitals (http://hospitals.webometrics.info)

The Ranking Web of Hospitals not only consist of evaluating the performance of the hospital activity, based solely on its production on the web, but also quantifies a wide range of activities different from the usual ones that measure the current generation of Bibliometric indicators, which focus only on the activities that generate the scientific elite. Hospital activity is defined by its multidimensionality, which is reflected in its web presence. Therefore, the best way to structure the ranking is through the combination of a group of indicators that measures all these different aspects. The ranking is updated twice a year (at the end of January and July) and provides all the information corresponding to some 16,000 hospitals as well as equivalent medical centers from all over the world.

10.12.4 Ranking Web of Research Centers (http://research.webometrics.info)

The Ranking Web of Research Centers is mainly responsible for measuring the volume, visibility, and impact of web pages published by research centers, not only placing special emphasis on scientific production but also taking into account the general information of the institution, its departments, research groups or support services, and the personnel work.

The ranking is updated twice a year (at the end of January and July) and already contains more than 8,000 research centers around the world. To compute the composite index (world ranking), it is compulsory to combine the normalized values; the positions based on these values are shown for informational purposes.

The Ranking Web of Repositories, Ranking Web of Hospitals, and the Ranking Web of Research Centers use the following four indicators that will be explained in a general way hereunder:

- *Size* (S): S is the total number of pages obtained from search engines.
- *Visibility* (V): V is the total number of external links received (Inlinks) by a website.
- *Rich files* (R): The following file types were selected after assessing their relevance in the academic and editorial environment and their volume in terms

of use with respect to other formats: Adobe Acrobat (.pdf), Adobe PostScript (.ps), Microsoft Word (.doc), and Microsoft PowerPoint (.ppt).

■ *Scholar* (Sc): GS supplies the number of articles and citations for each academic domain. The results from the GS database include articles, reports, and other associated material.

For the repositories, values related with altmetrics have been added. The values obtained from the references to contents of the repositories have been added to the visibility indicator in the following sources: Academia, Facebook, LinkedIn, Mendeley, ResearchGate, SlideShare, Twitter, Wikipedia, and YouTube (25%).

10.12.5 Ranking Web of Business School (http://business-schools.webometrics.info)

Even though Ranking Web of Business School appears on the website of the Cybermetrics Lab of the CSIC, the Ranking Web of Business Schools and institutions dedicated to the award of master of business administration degrees have been canceled and will not be updated in the future. The lack of transparency of these organizations when it comes to publishing information about their activities and achievements on the web makes it impossible to maintain this ranking.

10.12.6 Ranking Web of Researchers

Ranking Web of Researchers is a ranking that is currently in its beta phase and consists of classifying scientists based on their GS Public profiles. The lists only include the public profiles of scientists who voluntarily created the profiles in the GSC database. The value of the indicators is only valid for the marked collection date and is not updated automatically.

The rankings are world: top scientists (h >100); selected topics; Latin America; The Gulf; Europe; Asia, Africa, and Oceania; and Spain.

The ranking of scientists, which is based on their performance (h-index, citations) according to their public profiles of GSC, emerges to complement the information at an institutional level.

10.12.7 Worldwide Scientific Production in the Field of Webometrics (1997–2016)

For the purpose of completing this chapter and with the aim of providing current information about the state of production in Webometrics in the world, a query is made by introducing *Webometrics* in the article title, abstract, and keywords fields of the international Scopus database. The publication window will range from Ingwersen's first work in 1997 to the current year (2016) at the time of this writing. This is just a sample of how and in what state this field is.

A total of 350 papers have been published since the birth of Webometrics until today.

In Figure 10.5, the scientific production of Webometrics since it was named in 1997 until the last full year in Scopus (2016) can be observed.

Table 10.1 lists the authors worldwide who have published the most works about Webometrics over the years. In Table 10.2, the scientific journals which publish the most studies on Webometrics can be observed. The table includes the SJR 2015, the quartile journal rank in the discipline of library and information science, and the number of citations that these published works have obtained.

Table 10.3 lists the institutions which have published the largest amounts of studies about Webometrics.

In Figure 10.6, the countries that produce the most scientific works on Webometrics can be observed. Figure 10.7 presents the scientific disciplines that publish the most in the field of Webometrics.

It is important to point out that 2012, 2014, and 2015 were the years in which the most Webometrics studies were done. There were a smaller number of papers published when Webometrics emerged as a discipline around 1997. It is in the year 2004 when Webometrics began to take off with greater intensity.

As shown by the "Author" column in Table 10.1, Mike Thelwall, a leader of the Statistical Cybermetrics Research Group at the University of Wolverhampton, is the most published in this field; followed by Han Woo Park of Yeungnam University, Department of Media and Communication, Gyeongsan, South Korea;

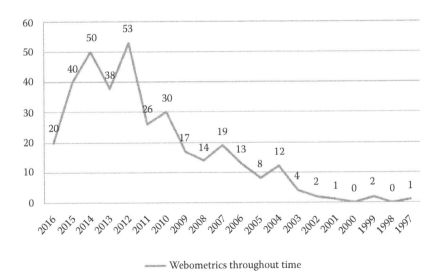

Webometrics throughout time

Figure 10.5 Worldwide scientific production in the field of Webometrics divided by years (1997–2016).

Table 10.1 Ten Most Productive Authors in the World in the Field of Webometrics

Author	Webometric Production	h-Index	Citations Received	Institution/Area	Country
Thelwall, M.	49	46	983	University of Wolverhampton, Statistical Cybermetrics Research Group, Wolverhampton	United Kingdom
Park, H. W.	36	21	481	Yeungnam University, Department of Media and Communication, Gyeongsan	South Korea
Aguillo, I. F.	23	15	428	CSIC, Cybermetrics Lab	Spain
Ortega, J. L.	17	12	290	CSIC, Cybermetrics Lab	Spain
Vaughan, L.	16	21	359	Western University, Faculty of Information and Media Studies, London	Canada
Orduña-Malea, E.	11	5	42	Polytechnic University of Valencia, EC3 Research Group	Spain
Danesh, F.	6	1	1	Ferdowsi University of Mashhad, Department of Library and Information Science, Mashhad	Iran
Holmberg, K.	6	9	69	University of Turku	Finland
Kousha, K.	6	14	81	University of Wolverhampton, Statistical Cybermetrics Research Group, Wolverhampton	United Kingdom
Buckley, K.	5	8	10	University of Wolverhampton, Statistical Cybermetrics Research Group, Wolverhampton	United Kingdom

Table 10.2   Scientific Journals with the Highest Number of Published Works Related to Webometrics

Journal	Webometric Production	SJR 2015	Quartile L&IS	Citations Obtained
Scientometrics	52	1.214	Q_1	885
Journal of the American Society for Information Science and Technology[a]	13	SJR 2014: 1.386	Q_1	422
Journal of Information Science	12	0.629	Q_1	228
Journal of Informetrics	12	1.803	Q_1	291
Aslib Proceedings New Information Perspectives[b]	9	SJR 2014: 0.57	Q_1	52
Cybermetrics	9	1.107	Q_1	89
Information Sciences and Technology	7	0.103	Q_4	0
Iranian Journal of Information Processing Management	7	0.1	Q_4	2
Journal of the Association for Information Science and Technology	7	1.601	Q_1	48
El Profesional de la Informacion	7	0.422	Q_2	37

Note: L&IS, library and information science.

[a] Formerly known as the *Journal of the American Society for Information Science and Technology*. Scopus coverage years: from 2000 to 2014 (coverage discontinued in Scopus). Continued as the *Journal of the Association for Information Science and Technology*.

[b] Formerly known as *Aslib Proceedings: New Information Perspectives*. Scopus coverage years: from 1976 to 1978, from 1982 to 1983, and from 1987 to 2014 (coverage discontinued in Scopus). Continued as *Aslib Journal of Information Management*.

Table 10.3 Institutions That Have the Highest Number of Works Published on Webometrics

Affiliation	Webometric Production
University of Wolverhampton	49
Yeungnam University	36
Western University	16
Spanish National Research Council (CSIC)	14
Polytechnic University of Valencia	12
University of Granada	12
CSIC–CCHS Institute of Information Studies on Science and Technology	8
Universidade Federal do Rio de Janeiro	6
CSIC–Centre for Human and Social Sciences	6
Isfahan University of Medical Sciences	6

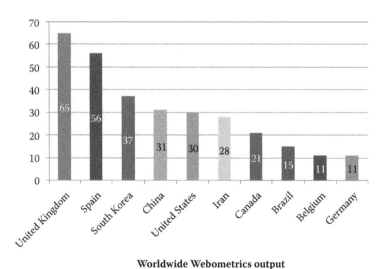

Worldwide Webometrics output

Figure 10.6 Countries with the largest amount of Webometrics production.

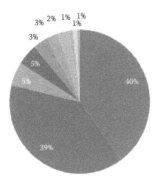

- Computer science
- Decision sciences
- Engineering
- Medicine
- Economics, econometrics, and finance
- Social sciences
- Mathematics
- Business, management, and accounting
- Earth and planetary sciences
- Environmental science

Figure 10.7   Webometric production sorted by subject area.

and Isidro F. Aguillo, leader of Cybermetrics Lab at the Spanish National Research Council (CSIC).

The most productive journal with published articles on Webometrics is *Scientometrics*, with 52 articles published in this area, followed by the *Journal of the American Society for Information Science and Technology* and the *Journal of Information Science*. These three were the first-quartile impact journals.

On the other hand, the University of Wolverhampton is the institution with the highest number of published works, followed by Yeungnam University and Western University. And the three Spanish institutions that publish the most works are the Spanish National Research Council (CSIC), Polytechnic University of Valencia, and University of Granada.

The country with the largest production in this field is the United Kingdom, with 65 papers; Spain, with 56; and South Korea, with 37.

The subject areas that are covered the most in Webometrics are computer science, social sciences, decision sciences, mathematics, and engineering.

References

Abraham, R.H. (1996). Webometry: Measuring the complexity of the World Wide Web. *Journal of New Paradigm Research*, 50:785–791.
Aguillo, I. (2012). La necesaria evolucion de la cibermetria. *Anuario ThinkEPI*, 6:119–122.

Almind, T.C., & Ingwersen, P. (1996). *Informetric Analysis on the World Wide Web: A Methodological Approach to "Internetometrics."* Copenhagen: Centre for Informetric Studies, Royal School of Library and Information Science (CIS Report 2).

Almind, T.C., & Ingwersen, P. (1997). Informetric analyses on the World Wide Web: Methodological approaches to Webometrics. *Journal of Documentation*, 53:404–426.

Baron, L., Tague-Sutcliffe, J., Kinnucan, M., & Carey, T. (1996). Labeled, typed links as cues when reading hypertext documents. *Journal of the American Society for Information Science*, 47:896–908.

Björneborn, L. (2004). *Small-World Link Structures across an Academic Web Space: A library and Information Science Approach.* Copenhagen: Department of Information Studies, Royal School of Library and Information Science.

Björneborn, L., & Ingwersen, P. (2004). Toward a basic framework for webometrics. *Journal of the American Society for Information Science and Technology*, 55:1216–1227.

Bossy, M.J. (1995). The last of the litter: "Netometrics." Solaris, 2 (special issue on "Les sciences de l'information: Bibliométrie, scientométrie, infométrie"). Presses Universitaires de Rennes.

Boutell, Th. (1994). Frequently asked questions about World-Wide Web. Electronic document.

Brin, S., & Page, L. (1998). The anatomy of a large-scale hypertextual Web search engine. Computer Networks and ISDN Systems, 30, 1–7.

Chakrabarti, S., Joshi, M.M., Punera, K., & Pennock, D.M. (2002). The structure of broad topics on the Web. Proceedings of the WWW2002 Conference, 251–262, 2002.

Cronin, B. (2001). Acknowledgement trends in the research literature of information science. *Journal of Documentation*, 57:427–433.

Dhyani, D., Bhowmick, S.S., & Ng, W.K. (2002). Web informetrics: Extending classical informetrics to the web. In: Proceedings of the 13th International Workshop on Database and Expert Systems Applications (DEXA'02). Washington, DC: IEEE Computer Society, 351–355.

Faba-Pérez, C., Guerrero-Bote, V., & Moya-Anegón, F. (2004). *Fundamentos y técnicas cibermétricas*. Mérida, Mexico: Junta de Extremadura.

García-Santiago, M.D. (2001). Topología de la información en la World Wide Web: Modelo experimental y bibliométrico en una red hipertextual nacional. Tesis Doctoral. Granada, Spain: Universidad, Departamento de Biblioteconomía y Documentación.

Garfield, E. (1955). Citation indexes to science: A new dimension in documentation through association of ideas. *Science*, 122:108–111.

Heylighen, F., & Bollen, J. (1996). The World-Wide Web as a super-brain: From metaphor to model. In: R. Trappl (Ed.) *Cybernetics and Systems '96*. Vienna: Austrian Society for Cybernetics, 917–922.

Ingwersen, P. (1998). The calculation of Web impact factors. *Journal of Documentation*, 54:236–243.

Kim, H.J. (2000). Motivations for hyperlinking in scholarly electronic articles: A qualitative study. *Journal of the American Society for Information Science*, 51:887–899.

Larson, R.R. (1996). Bibliometrics of the World Wide Web: An exploratory analysis of the intellectual structure of Cyberspace. Proceedings of the 59th ASIS Annual Meeting, Baltimore, Maryland. Medford, NJ: Learned Information Inc./ASIS, 71–78.

Lascurain, M.L. (2006) La evaluación de la actividad científica mediante indicadores bibliométricos. *Boletín Bibliotecas*, 24:1–12.

McKiernan, G. (1996). CitedSites (sm): Citation indexing of web resources. http://www
.public.iastate.edu/~CYBERSTACKS/Cited.htm [Consulted: 12/14/2016].

Orduña-Malea, E. (2012). Propuesta de un modelo de análisis redinformétrico multinivel
para el estudio sistémico de las universidades españolas. Valencia [doctoral thesis].

Orduña-Malea, E., & Ontalba-Ruipérez, J.-A. (2013). Proposal for a multilevel university
cybermetric analysis model. *Scientometrics*, 95:863–884.

Orduña-Malea, E., & Aguillo, I.F. (2014). *Cibermetría: Midiendo el espacio red*. Barcelona:
UOC (EPI Scholar).

Smith, A.G. (1999). The impact of web sites: A comparison between Australasia and Latin
America. In: Proceedings of INFO'99, Congreso Internacional de Informacion,
Havana, 4–8 October.

Thelwall, M. (2003). What is this link doing here? Beginning a fine-grained process of
identifying reasons for academic hyperlink creation. *Information Research*, 8(3) paper
no. 151.

Thelwall, M., Vaughan, L., & Björneborn, L. (2005). Webometrics. *Annual Review of
Information Science and Technology*, 35:81–135.

Thelwall, M., & Sud, P. (2012). Webometric research with the Bing Search API 2.0.
Journal of Informetrics, 6:44–52.

Velasco-Gatón, B., Eiros-Bouza, J., Pinilla-Sánchez, J., & San Román-Calvar, J. (2012).
La utilización de los indicadores bibliométricos para evaluar la actividad investiga-
dora. *Aula Abierta*, 40:75–84.

Chapter 11

U-Multirank: Data Analytics and Scientometrics

Frans Kaiser[1] and Nadine Zeeman[2]

[1]*Center for Higher Education Policy Studies, University of Twente, The Netherlands*
[2]*Center for Higher Education, TU Dortmund*

Contents

11.1   Introduction

U-Multirank (UMR) is a European-based, global transparency tool in the realm of higher education and research. Since 2014, UMR has published an annual web tool in which a variety of users can compare higher education institutions (HEIs) and their programs in a user-driven way. Since UMR uses multiple sources (from institutional questionnaires, student questionnaires to the well-known citation indexes), we present UMR from the perspective of Data Analytics and Scientometrics. In short, we explain how the data for UMR are collected, the different ways that the data are analyzed, and how the results are shared with the audiences of UMR.

Before describing the data sources, data verification, and the data presentation, we will describe the kind of information that UMR comprises on HEIs. Section 11.2 therefore addresses the dimensions and the indicators used in UMR.

In Section 11.3, we describe the various data sources that are used to collect the data. UMR retrieves data from various sources, including online surveys among participating institutions, student surveys, existing databases on publications and citations, patent databases, national administrative databases, and international databases. The description will focus primarily on the institutional level.

The participating institutions provide their data via online questionnaires on the institutional and subject level. Section 11.4 addresses how the self-reported data of the HEIs are verified.

Sections 11.5 and 11.6 focus on the way that the results are presented and communicated. We will describe how the interactive, user-driven, and multidimensional web tool has been developed and designed. In addition, we will focus on a distinguishing characteristic, that is, the categorization of results in *rank groups*.

11.1.1 Overview of UMR

UMR is a transparency tool in which quantitative information on the performance of HEIs is presented in a comparative way. UMR is also a research and development project in which the existing tool was developed and designed. The project started with a feasibility study in 2009 and was continued as a development project in 2013.

The reason for starting the projects was to create an alternative to the existing global university rankings and by doing so enhancing the transparency of the performance of HEIs in the European Higher Education Area and beyond.

The critique on the existing rankings focused on a variety of issues such as the use of a narrow range of dimensions, the use of composite overall indicators, the use of league tables, the problem of unspecified target groups, the problem of ignoring diversity within HEIs, and the problem of poorly specified and volatile methodologies. The assessment of these problems and issues led to five design principles that were used to develop a new ranking (Federkeil et al., 2012c, pp. 85–86; Federkeil et al., 2012d, pp. 39–70).

11.1.2 Five Design Principles

UMR has been developed based on a number of design principles (Federkeil et al., 2012c, pp. 86–87):

1. *User-driven*: Rankings should be based on the different interests and priorities that are determined by its users. Users are the most important actors in the design and the application or rankings. It is therefore necessary that users and stakeholders have the opportunity to design their own rankings.
2. *Multidimensionality*: HEIs have more than one purpose and mission and undertake a different set of activities. These activities include activities in the five major dimensions that we have identified: teaching and learning, research, knowledge transfer, regional engagement, and internationalization. Rankings should not only focus on one function, but should also reflect this variety of functions of HEIs.
3. *Comparability*: HEIs and their programs should only be compared when their purposes and activity profiles are to a large extent similar. It does not

make sense to compare HEIs and programs in rankings that have very different functions and objectives.

4. *Multilevel nature of higher education*: Rankings should reflect the multilevel nature of higher education. In most cases, HEIs are a combination of faculties, departments, and programs that differ in strength. An aggregated institutional ranking disguises this reality and does not produce the information that is being valued by important stakeholders, such as (potential) students, employers and professional organizations, and academic staff. It is relevant to create a ranking that enables users to compare similar institutions at the institutional level and at the level of the disciplinary departments and subjects in which they are active.

5. *Methodological soundness*: The ranking should avoid methodological mistakes. These include, for example, composite indicators, the production of league tables, and the denial of context.

Based on these design principles, the transparency tool was developed and tested, leading to the launch of UMR in 2014.

11.2 Dimensions and Indicators

11.2.1 Overview of the Dimensions

For both the institutional and the subject ranking, we identified five dimensions. The teaching and learning dimension reflects a core activity in most HEIs. This core activity consists of all the processes concerned with the dissemination of knowledge, skills, and values to students. The second dimension of research is focusing on activities of HEIs on basic and applied research. The focus on these elements is justified by the growing diversity of the research missions of HEIs, for example, the traditional research universities and the more vocational-oriented institutions (university colleges, universities of applied sciences, etc.). The third dimension *knowledge transfer* is concerned with the activities of HEIs to make science output available for economic, social, and cultural development of communities, for example, nations and regions. The international orientation dimension reflects the activities of HEIs with regard to developing and promoting international academic staff and student mobility, developing and increasing international cooperation, and developing and expanding international competition. The last dimension *regional engagement* compromises the activities of HEIs to play a role in the process of creating favorable conditions for a region to develop and to thrive. This activity is part of the third mission of an institution (Federkeil et al., 2012b, pp. 98–120).

11.2.2 Institutional and Subject Ranking Dimensions and Indicators

UMR provides a multidimensional ranking both on the institutional and on the subject levels. UMR therefore offers two levels of rankings. For the institutional and the subject ranking, five dimensions have been identified, and for each of these dimensions, indicators have been selected to measure the performance of HEIs. The choice of indicators was highly stakeholder driven. The selection has been based on the earlier UMR feasibility study and adapted to include suggestions made by various stakeholders and the UMR Advisory Board (Federkeil et al., 2012b, p. 97; U-Multirank, 2015a, p. 10).

11.2.2.1 Institutional Ranking Indicators

Table 11.1 provides an overview of the UMR institutional ranking indicators for each of the five dimensions. UMR currently provides a comprehensive set of 35 ranking indicators. The table shows the dimensions, the indicators, the definition of the indicators as well as the rationale for each of the indicators (U-Multirank, 2015a; U-Multirank, 2016a, pp. 7–37)

11.2.2.2 Subject Ranking Indicators

Since the main focus here is on the institutional ranking indicators, we will present the subject ranking indicators in a descriptive manner. The subject ranking indicators focus on particular academic disciplines or groups of programs and are an important criterion for the success of UMR. The subject rankings are very relevant for many of the users of UMR. This is especially the case for students who benefit from the option to compare universities in the subject that they intend to study. In addition, it is also beneficial for academics who are interested to compare their department with other departments or programs in their field (Van Vught & Ziegele, 2011, pp. 35, 41).

For the latest UMR ranking of 2017, some of the subjects have been updated for the first time. In addition, new subjects have been added. The following subjects have been included in the 2017 ranking: business studies (update), economics (new), computer science and engineering (update), mechanical engineering (update), electrical engineering (update), chemical engineering (new), industrial engineering/production (new), and civil engineering (new). The number of indicators of the subject rankings varies between the subjects from 26 to 38 indicators.

Table 11.1   U-Multirank Institutional Ranking Indicators

Dimension	Indicator	Definition	Rationale
Teaching and learning	Bachelor graduation rate	The percentage of new entrants that successfully completed their bachelor program.	The graduation rate shows how well the university's programs are organized and reflects the effectiveness of its teaching.
	Masters graduation rate	The percentage of new entrants that successfully completed their master program.	The graduation rate shows how well the university's programs are organized and reflects the effectiveness of its teaching.
	Graduating on time (bachelors)	The percentage of graduates that graduated within the time expected (normative time) for their bachelor program.	The time to degree reflects how well the university's programs are organized and shows the effectiveness of its teaching.
	Graduating on time (masters)	The percentage of graduates that graduated within the time expected (normative time) for their master's program.	The time to degree reflects how well the university's programs are organized and shows the effectiveness of its teaching.
Research	Citation rate	The average number of times that the university's research publications (over the period 2011–2014) are cited in other research; adjusted (normalized) at the global level to take into account differences in publication years and to allow for differences in citation customs across academic fields.	Indicator of the scientific impact of research outputs within international scientific communities. This normalization aims to correct for differences in citation characteristics between publications from different fields and different years.

(Continued)

Table 11.1 (Continued) U-Multirank Institutional Ranking Indicators

Dimension	Indicator	Definition	Rationale
	Research publications (absolute numbers)	The number of university's research publications (indexed in the WoS Core Collections database), where at least one author is affiliated to the source university or higher education institution.	The number of publications in academic journals is a measure of the institution's research activity and its capability in producing research publications at the international level.
	Research publications (size normalized)	The number of research publications (indexed in the WoS database), where at least one author is affiliated to the university (relative to the number of students).	The number of publications in academic journals is a measure of the institution's research activity and its capability in producing research publications at the international level, correcting for the size of the institution (approximated by student enrolled).
	External research income	Revenue for research that is not part of a core (or base) grant received from the government. Includes research grants from national and international funding agencies, research councils, research foundations, charities, and other nonprofit organizations. Measured in € 1,000s, using PPP. Expressed per fte academic staff.	The indicator expresses the institution's success in attracting grants in national and international competitive, peer-reviewed programs. This reflects the quality of an institution's research.

(Continued)

Table 11.1 (Continued) U-Multirank Institutional Ranking Indicators

Dimension	Indicator	Definition	Rationale
	Art related output	The number of scholarly outputs in the creative and performing arts, relative to the fte number of academic staff.	This measure recognizes outputs other than research publications and reflects all tangible research-based outputs such as musical compositions, designs, artifacts, software, etc.
	Top cited publications	The proportion of the university's research publications that, compared to other publications in the same field and in the same year, belong to the top 10% most frequently cited worldwide.	This is a measure of international research excellence. Departments with well over 10% of their publications in the top percentile of frequently cited articles worldwide are among the top research institutes worldwide.
	Interdisciplinary publications	Extent to which reference lists of university's publications reflect cited publications in journals from different scientific disciplines	The more a publication refers to publications belonging to different fields of science and the larger the distance between these fields, the higher the degree of interdisciplinarity. More interdisciplinarity signals excellence, given that the frontiers of research are often at the edge of discipline.
	Postdoc positions	The number of postdoc positions relative to the number of academic staff (headcount).	As post doc positions are often externally (and competitively) funded, an institution with more postdoc positions is more likely to have a higher research quality.

(Continued)

Table 11.1 (Continued) U-Multirank Institutional Ranking Indicators

Dimension	Indicator	Definition	Rationale
	Professional publications	The number of professional publications per fte academic staff. Professional publications are all publications published in journals, books, and other media that are addressed to a professional audience and that can be traced bibliographically. These publications are not peer reviewed as in the category academic publications.	Most of the indicators in the research dimension focus on academic publications. This indicator introduces a wider audience for research activities: professional communities. This indicator allows HEIs that are focused on professional audiences to become more visible with their research activities and outputs.
Knowledge transfer	Copublications with industrial partners	The percentage of the university's research publications that list an author affiliate with an address referring to a for-profit business enterprise or private sector R&D unit (excludes for-profit hospitals and education organizations).	The more research is carried out with external partners, the more likely it is that knowledge transfer takes place between academia and business.
	Income from private sources	Research revenues and knowledge transfer revenues from private sources (incl. not-for-profit organizations), excluding tuition fees. Measured in €1,000s using PPP. Expressed per fte academic staff.	The degree to which research is funded by external, private organizations reflects aspects of its research quality—most notably its success in attracting funding and research contracts from end-user sources.

(Continued)

Table 11.1 **(Continued)** U-Multirank Institutional Ranking Indicators

Dimension	Indicator	Definition	Rationale
	Patents awarded (absolute numbers)	The number of patents granted at the USPTO and/or the EPO for patents applied between 2003 and 2012.	The number of patents is an established measure of technology transfer as it indicates the degree to which discoveries and inventions made in academic institutions may be transferred to economic actors for further industrial/commercial development.
	Patents awarded (size normalized)	The number of patents granted at the USPTO and/or the EPO for patents applied between 2003 and 2012 (per 1,000 students).	The number of patents is an established measure of technology transfer as it indicates the degree to which discoveries and inventions made in academic institutions may be transferred to economic actors for further industrial/commercial development.
	Industry copatents	The percentage of the number of patents assigned to (inventors working in) the university over the period 2003–2012, which were coapplied with at least one applicant from the industry.	If the university applies for a patent with a private firm, this reflects that it shares its knowledge with external partners and shows the extent to which it is willing to share its technological inventions for further commercial development.

(Continued)

Table 11.1 (Continued) U-Multirank Institutional Ranking Indicators

Dimension	Indicator	Definition	Rationale
	Spin-offs	The number of spin-offs (i.e., firms established on the basis of a formal knowledge transfer arrangement between the institution and the firm) recently created by the institution (per 1,000 fte academic staff).	A new firm that is based on knowledge created in a university signals a successful case of knowledge transfer from academia to industry.
	Publications cited in patents	The percentage of the university's research publications that were mentioned in the reference list of at least one international patent (as included in the PATSTAT database).	This indicator reflects the technological relevance of scientific research at the university, in the sense that it explicitly contributed, in some way, to the development of patented technologies.
	Income from continuous professional development	The percentage of the university's total revenues that is generated from activities delivering Continuous Professional Development courses and training.	When a university is very active in providing continuing education courses to companies and private individuals, it transfers knowledge to its environment.

(Continued)

Table 11.1 (Continued) U-Multirank Institutional Ranking Indicators

Dimension	Indicator	Definition	Rationale
	Graduate companies	The number of companies newly founded by graduates per 1,000 graduates.	The number of companies newly founded by graduates refers to any company that graduates of the higher education institution have founded. Any type of registered company (for-profit/not-for-profit/small/large; manufacturing/service/consultancy) may be reported here. The information will be used to assess the entrepreneurial orientation of the higher education's graduates. Graduates as the denominator of the ratio is a logical choice as the number of graduate companies is likely to be related to the total number of graduates.
	Strategic research partnerships	The number of strategic partnerships per fte academic staff. A strategic partnership is a formal alliance between the higher education institution (or part of it) and one or more external organizations with which a long-term agreement is reached for sharing of physical and/or intellectual resources in the achievement of defined common goals. The focus lies here on agreements referring to research and knowledge exchange activities.	Strategic partnerships with a focus on research indicate the commitment of HEI and its environment to engage in research collaboration. This collaboration is likely to focus on applied research activities. Academic staff as the denominator is an adequate choice as academic staff has to engage in these activities. More staff is likely to be related to more partnerships.

(Continued)

Table 11.1 (Continued) U-Multirank Institutional Ranking Indicators

Dimension	Indicator	Definition	Rationale
International orientation	Foreign language bachelor programs	The percentage of bachelor programs that are offered in a foreign language.	Offering degree programs in a foreign language signals the commitment of the university to welcome foreign students and to prepare its students for working in an international environment.
	Foreign language master programs	The percentage of master's programs that are offered in a foreign language.	Offering masters programs in a foreign language testifies the commitment of the university to welcome foreign students and to prepare its students for working in an international environment.
	Student mobility	A composite of international incoming exchange students, outgoing exchange students, and students in international joint degree programs.	Having an international student body and offering students the opportunity to do part of their degree abroad signals the international orientation of the university.
	International academic staff	The percentage of academic staff (on a headcount basis) with foreign citizenship.	Having an international academic staff reflects the international orientation of the university and its attractiveness as an employer for foreign academics.

(Continued)

Table 11.1 (Continued) U-Multirank Institutional Ranking Indicators

Dimension	Indicator	Definition	Rationale
	International joint publications	The percentage of the university's research publications that list at least one affiliate author's address located in another country.	The number of international joint publications reflects the degree to which a university's research is connected to international networks.
	International doctorate degrees	The percentage of doctorate degrees that are awarded to international doctorate candidates.	The number of doctorate degrees awarded to international candidates reflects the international orientation of an institution.
Regional engagement	Bachelor graduates working in the region	The percentage of bachelor graduates who found their first job (after graduation) in the region where the university is located.	If a relatively large number of an institution's graduates is working in the region, this reflects strong linkages between the university and its regional partners.
	Master graduates working in the region	The percentage of masters graduates who found their first job (after graduation) in the region where the university is located.	If a relatively large number of an institution's graduates is working in the region, this reflects strong linkages between the university and its regional partners.

(Continued)

Table 11.1 **(Continued)** U-Multirank Institutional Ranking Indicators

Dimension	Indicator	Definition	Rationale
	Student internships in the region	Out of all the university's students who did an internship, the percentage where the internship was with a company or organization located in the region.	Internships of students in regional enterprises are a means to build cooperations with regional partners and connect students to the local labor market.
	Regional joint publications	The percentage of the university's research publications that list at least one coauthor with an affiliate address located in the same spatial region (within a distance of 50 km).	Copublications with authors located elsewhere in the institution's geographical region are a reflection of regional linkages between the university and regional partners.
	Income from regional sources	The proportion of external research revenues — apart from government or local authority core/recurrent grants — that comes from regional sources (i.e., industry, private organizations, charities).	A high proportion of income from regional/local sources indicates a more intense relationship between the university and the region.
	Strategic research partnerships in the region	The number of strategic research partnerships with partners in the region as a percentage of the total number of strategic research partnerships.	A higher education institution that finds most of its partners for research activities in the region is most likely to be engaged in the region.

11.2.3 Mapping Indicators: Comparing Like with Like

It was clear at the very start of the project that UMR should be a tool that allows for transparency regarding the diversity of HEIs. UMR includes HEIs with diverse missions, structures, and profiles, which makes it necessary to specify this diversity (Federkeil et al., 2012a, p. 168).

The UMR web tool follows a like-with-like approach to enable users to compare HEIs with similar and hence comparable institutional profiles. HEIs with broadly similar profiles have to be identified by the user, based on indicators of their particular characteristics and activities. The UMR web tool offers a facility for users to define these profiles based on their own needs and interest. The UMR *Compare like with like* function presents 12 criteria to identify subsets of comparable institutions which work as filters and determine the sample of institutions to be included in the personalized ranking. This *like-with-like* selection is based on mapping indicators such as size, scope, age, or activity (U-Multirank, 2015a, pp. 10, 15).

The mapping indicators make it possible to create multiple rankings with comparable institutions and to show specific performance profiles. The user can define the institutional profile that they are interested in by selecting the indicators that they deem important out of the six dimensions. These six dimensions are teaching and learning, research, knowledge transfer, international orientation, regional engagement, and general section (Federkeil et al., 2012a, p. 168; Van Vught & Ziegele, 2012, p. 179 ; U-Multirank, 2016a, pp. 90–100). The mapping indicators, thus the criteria that the user can select to create a sample of institutions to be compared on an alike basis, are shown in the following list:

1. Teaching and learning
   a. *Expenditure on teaching*: Percentage of total institutional expenditure dedicated to teaching activities.
   b. *Graduate students*: The number of higher degrees (master and PhD) awarded as a percentage of total number of degrees awarded.
   c. *Level of study*: The degree levels at which the institution awards degrees.
   d. *Specialization*: For those HEIs that have a specialized character, the largest subject field in which the HEI has specialized (in terms of graduates) is presented.
2. Research
   a. *Expenditure on research*: The percentage of expenditure allocated to research activities.
3. Knowledge transfer
   a. *Income from private sources*: The total amount of external research income and income from knowledge transfer from private sources as a percentage of total income of institution.

4. International orientation
    a. *Foreign degree-seeking students*: The percentage of degree-seeking students with foreign qualifying diplomas.
5. Regional engagement
    a. *New entrants from the region*: The percentage of new entrants to bachelor programs coming from the region in which the institution is located.
6. General
    a. *Size*: The size of the institution in terms of the number of students enrolled.
    b. *Legal status*: The public/private character of the institution.
    c. *Age*: The age of the institution, based on the founding year of the oldest part of the institution.

11.2.4 Contextualizing and Enriching the Results: Institutional Descriptive Indicators

UMR also provides some additional information on the university such as its location, the self-reported number of academic publications that it has produced (per fte academic staff) according to its own counts, the rate of unemployment of its graduates (at bachelor and/or master and/or long-first-degree level; 18 months after graduation), and the profiles of degree programs (U-Multirank, 2015a, p. 10; U-Multirank, 2016a, pp. 38–45).

It is important to briefly explain why the indicator on graduate employment is included as a descriptive indicator and not as a ranking indicator. The indicator is a problematic indicator in that is has proven to be difficult to retrieve reliable and international comparable data, as the feasibility study has demonstrated. Still, the stakeholders put forward that it is a highly relevant indicator. For this reason, the data on graduate employment are not presented as a ranking indicator, but instead, the data for a limited number of institutions are presented as a descriptive indicator (U-Multirank, 2015a, p. 23).

We have included the following institutional descriptive indicators:

1. *Graduation rate long first degree*: The percentage of new entrants that successfully completed their long-first-degree program
2. *Graduating on time (long first degree)*: The percentage of graduates that graduated within the time expected (normative time) for their long-first-degree program
3. *Relative bachelor graduate unemployment*: The percentage of 2013 bachelor graduates who were not in education, employment, or training (NEET) in 2015 (some 18 months later)
4. *Relative master graduate unemployment*: The percentage of 2013 master graduates who were NEET) in 2015 (some 18 months later)
5. *Relative graduate unemployment long first degree*: The percentage of 2013 long first graduates who were NEET in 2015 (some 18 months later)

6. *Publication output*: The number of all research publications included in the institution's publication databases, where at least one author is affiliated to the institution (per fte academic staff)
7. *Foreign language long-first-degree programs*: The percentage of long-first-degree programs that are offered in a foreign language

In the UMR 2017 publication, we introduced four new descriptive indicators on gender balance to include a social dimension. Gender balance has been included because it is regarded as an important characteristic of the learning environment. These four new descriptive indicators are as follows:

1. *Female bachelor students*: The number of female students enrolled in bachelor programs as a percentage of total enrollment in bachelor programs
2. *Female master students*: The number of female students enrolled in master programs as a percentage of total enrollment in master programs
3. *Female academic staff*: The number of female academic staff as a percentage of total number of academic staff
4. *Female professors*: The number of female professors as a percentage of total number of professors

11.2.5 Student Survey

The subject-based rankings include indicators that are derived from a student survey. One of the purposes of UMR is to help prospective and increasingly mobile students to make an informed choice about a university. For them, the assessment of the learning experience by current students of institutions provides a unique peer perspective. The main instrument for measuring the student satisfaction is an online survey that includes all the subjects, asking the opinion of students on various aspects about their degree program. The online survey consists of open-ended and closed-ended questions. The main focus of the student survey is to assess the teaching and learning experience and the faculties of the institution. The indicators derived from the student survey reflect the different aspects of the learning experiences of students. Since they refer to a particular study program, they are being used only in the subject-based ranking (Callaert et al., 2012, p. 133; U-Multirank, 2015a, p. 9).

11.3 Data Sources

11.3.1 UMR Surveys

Universities that participate in UMR provide data for the institution as a whole, as well as for the departments offering degree programs (if any) related to the selected academic subjects covered in UMR. In addition, students are invited to participate in the student survey sent out through the universities. UMR therefore relies to a large extent on self-reported data at the institutional and subject-based levels. The self-reported data are directly collected from the HEIs and their students via three online surveys: the institutional questionnaire, the subject questionnaire, and the student survey (Callaert et al., 2012, p. 131; U-Multirank, 2015a, p. 9).

11.3.1.1 Institutional Questionnaire

The institutional questionnaire collects data on the performance at the level of the institution. The most recent institutional questionnaire used for the UMR 2017 ranking is divided into the following categories:

1. *General information*: Information of the contact person, legal name of the institution, type of institution, public/private character, foundation year, multisite, and university hospital
2. *Programs and students*: Degree programs offered, (total) enrollment, international students, and students in internships
3. *Graduates*: Graduates by level of program and educational field, graduate employment, and graduates working in the region
4. *Academic staff*: Fte and headcount, international staff, and postdocs
5. *Revenues:* total revenues, tuition fees, and external research revenues coming from international, national, and regional sources
6. *Expenditure*: Total expenditure and expenditure by activity (research, teaching, knowledge transfer, and other)
7. *Research and knowledge transfer*: Publications, art-related outputs, graduate founded companies, start-up firms, and strategic partnerships

11.3.1.2 Subject Questionnaires

The subject questionnaires include information on individual departments/faculties and their programs. On the department level, information is collected on staff and PhD, income, and students. On the level of individual study programs, data are collected on the programs offered, students enrolled, admission restrictions, tuition fees, work experience integrated in the programs, credits offered for community service-learning/social activities, the international orientation of the programs, the number of degrees issued (graduates), and graduate employment.

11.3.1.3 Student Survey

The inclusion of student satisfaction indicators on the subject level is a unique feature of UMR and highly appreciated by students and student organizations. On the one hand, this survey provides HEIs feedback given by their students on their degree programs offered. HEIs can thus immediately use the UMR findings for internal management issues. On the other hand, many HEIs are already carrying out their own student surveys, and in some countries, there are comprehensive national student surveys. In these institutions or countries, the implementation of an additional student survey is a major challenge, especially in terms of willingness of students to respond. Therefore, it is desirable to use existing student surveys as much as possible for UMR to avoid unnecessary doubling of surveys.

11.3.2 Publicly Available Data Sources

UMR strives to continuously reduce the burden on institutions in terms of data delivery. Instruments are being reviewed annually, the necessity of all data points in the questionnaires is being reconsidered every year, and the continuous refinement of data definitions based on institutional feedback is crucial. In this context, UMR continues to explore options to use publicly available data, from national and international data sources as well as from national student surveys (U-Multirank, 2015a, pp. 6, 22).

11.3.2.1 National Administrative Data Sources

National administrative data sources provide detailed information on HEIs which is missing at the international level on the system and/or subject level of HEI. There are limited valid comparative data available at the international level. This detailed information at the national level is offered by national ministries, national statistical bodies, and/or accreditation agencies. UMR has identified and explored national administrative data sources to use their data to reduce the burden on institutions in terms of data delivery. The countries covered are the United States, the United Kingdom, and Spain.

11.3.2.1.1 Integrated Postsecondary Education Data System

The Integrated Postsecondary Education Data System (IPEDS) is a system of inter-related surveys that are conducted annually by the U.S. Department of Education's National Center for Education Statistics. IPEDS collects data from institutions that participate in federal student aid programs. The institutions include, inter alia, (research) universities, state colleges, and technical and vocational institutions. These institutions need to provide data on institutional characteristics, enrollment, program completions (degrees and certificates conferred), graduation rates (student persistence and success), faculty and staff (human resources), finances (fiscal resources), institutional prices, and student financial aid (National Center for Education Statistics, n.d.).

UMR has prefilled the institutional questionnaire with data retrieved from IPEDS. The institutional questionnaires have been prefilled only with those data elements from IPEDS that correspond with the UMR definitions. The data of the IPEDS database are also used to collect data for the four new indicators on gender balance.

11.3.2.1.2 Higher Education Statistics Agency

The UK Higher Education Statistics Agency (HESA) publishes data on all elements of the higher education sector in the United Kingdom. HESA includes data about students in higher education, academic and nonacademic higher education staff, finances, public engagement, and commercial enterprises. The data are collected from universities, higher education colleges, and other specialist providers of higher education (Higher Education Statistics Agency, n.d.).

UMR has prefilled the institutional questionnaire with the data provided by HESA. The institutional questionnaires have been prefilled only with those data elements from HESA that correspond with the UMR definitions.

11.3.2.1.3 Spain

The data for the Spanish HEIs are jointly collected for their national ranking (Ranking CYD) as for UMR. In Spain, they organize their own surveys which are identical to the UMR surveys. The partner of UMR and organization responsible for the national ranking and the UMR data collection is the Foundation for Knowledge and Development (Fundación CYD). Fundación CYD coordinates the relationship between the Spanish universities and UMR. It collects the data at the level of the subject areas from those universities, as well as some data relating to the universities as a whole (U-Multirank, 2016c).

11.3.2.1.4 National Student Survey: Norway

A major advantage with regard to the use of existing national student surveys has been made on Norway. UMR and the Norwegian Agency for Quality Assurance in Education (NOKUT) have agreed on using the data of the national Norwegian student survey (Studiebarometeret) for UMR. NOKUT supplies UMR with the data and results from the annual national student survey so that the Norwegian HEIs do not have to carry out UMR's student survey. UMR could make use of this existing student survey because the Norwegian student survey is very similar to the UMR student survey. In the Studiebarometeret, Norwegian students can express their opinion on a wide range of dimensions of their study program. These include the quality of teaching and learning, the workload of students, and the career opportunities (U-Multirank, 2015a, p. 12; U-Multirank—International ranking of universities and university colleges, 2013).

11.3.2.2 International Data Sources

Next to the national administrative data sources, international data sources have also been explored. Again, this has been done to reduce the administrative burden for the institutions concerning the collection of data.

11.3.2.2.1 European Tertiary Education Register

The European Tertiary Education Register (ETER) is a database that comprises basic information on HEIs in Europe. They collect data on the basis characteristics and geographic positions of the institutions, and they provide data on students, educational activities (fields of education), staff, finances (income and expenditure), and research activities at the institutional level. This database, for which the data come from national statistical agencies, offers a unique potential base for comparison and research of European HEIs (European Tertiary Education Register, n.d.).

The data of ETER are being used to collect data for the mapping indicators *legal status* and *age* and to obtain qualitative descriptive information such as the address, country, and website of the institution. In addition, the data of the ETER database are also used to collect data for the four new indicators on gender balance to take into account the social dimension.

11.3.2.2.2 World Higher Education Database

The World Higher Education Database (WHED) is an online database provided by the International Association of Universities. The online database includes comprehensive information on higher education systems and credentials in 184 countries and over 18,000 HEIs. The data on systems and credentials compromise information on, inter alia, the different stages of study, educational level and entrance requirements, and main credentials. The data on HEIs include information on, inter alia, divisions, degrees offered, and student and academic staff numbers (International Association of Universities, 2016).

The data of the WHED database are used to collect data for the mapping indicators *scope*, *level of study*, and *age*. It is also used to obtain qualitative descriptive information such as the address, country, and website of the institution.

11.3.2.2.3 PATSTAT

Patent data that are used to calculate the patent-related indicators (publications cited in patents, patents awarded, and industry copatents) are derived from the PATSTAT database. PATSTAT is a database produced by the EPO that contains bibliographical data relating to more than 90 million patent documents from more than 100 leading industrialized and developing countries.

Patent publications usually contain references to other patents and sometimes also to other *nonpatent* literature sources. A major part of these nonpatent references (NPRs) is citations to scholarly publications published in WoS-indexed sources. The NPRs are the so-called front-page citations. These citations are mainly provided by the patent applicant(s) or by the patent examiner(s) during the search and examination phases of the patent application process.

The citing patents were clustered by using the *simple patent family* concept— that is, groupings of patent publications containing all equivalent, in legal sense, patent documents. A simple patent family therefore addresses one single *invention*. Each patent family contains at least one EP patent (published by EPO) or a WO patent (published by the World Intellectual Property Organization) and at least one patent published by USPTO. All NPRs within each family were deduplicated. Each NPR is therefore counted only once per family. The NPRs were matched against the bibliographical records in the WoS. Our current information indicates that the majority of the WoS records are identified (Federkeil, Van Vught and Westerheijden, 2012e, p. 34; U-Multirank, n.d.).

The patent database used to collect the NPRs is the spring 2014 version of the EPO Worldwide Patent Statistical Database (PATSTAT). The Centre for Science and Technology Studies (CWTS) operates on an EPO-licensed version of PATSTAT.

11.3.2.2.4 Bibliometric Data

All Bibliometric scores are based on information extracted from publications that are indexed in the WoS Core Collection database (SCIE, SSCI, and AHCI). CWTS operates this WoS database under a commercial license agreement with Thomson Reuters.

The WoS contains some 12,000 active sources, both peer-reviewed scholarly journals and conference proceedings. The underlying bibliographic information relates to publications classified as *research article* and *review article*. The WoS database is incomplete (there are many thousands more science journals worldwide), and it is biased in favor of the English language. Hence, there will always be missing publications. WoS-based Bibliometric data are never comprehensive and fully accurate; scores are therefore always estimates with a margin of statistical error.

Nonetheless, the WoS is currently one of the two best sources, covering worldwide science across all disciplines. The only possible alternative database, Elsevier's Scopus database, has more or less the same features. All in all, one may expect comparable Bibliometric results from both databases, especially at higher aggregate levels.

All the Bibliometric indicators presented in this section are either fully or partially derived from preexisting generally available indicators or based on prior CWTS ideas or research that occurred outside the UMR project. In some cases, the

Bibliometric scores on these indicators were derived from prior CWTS-developed data processing routines or computational algorithms or modified/upgraded versions thereof. The WoS-based Bibliometric scores relate to the publication years 2011 up to and including 2014, as a single measurement window, with the exception of the *Patent citations to research publications* metrics.

The Bibliometric indicators in UMR are closely related to those in the Leiden Ranking (LR). The main difference between both is the fact that LR is based on WoS-indexed *core research publications* in international peer-reviewed scientific journals. Publications in other WoS-indexed sources (national scientific journals, trade journals, and popular magazines) are not included. The same applies to research publications in languages other than English. Also, publications in journals that are not well connected, in terms of citation links, to other journals are left out (these are mainly, but not exclusively, journals in arts and humanities fields of science).* In contrast, UMR includes all WoS-indexed sources and publications (although for some indicators A&H publications are left out; see in the following paragraphs).

The Bibliometric indicators fall into two groups, depending on the counting scheme (fractional or full) and the coverage of the arts and humanities research publications (included or excluded). The following indicators use full counting and include the arts and humanities: research publication output, international copublications, regional copublications, and copublications with industrial partners. Three of the other indicators exclude arts and humanities publications and use a fractional counting scheme: interdisciplinary research score, mean normalized citation score, and frequently cited publications. Finally, patent citations to research publications also exclude arts and humanities publications, but it uses full counting. We refer to the recent publication by Waltman and Van Eck (2015; http://arxiv.org/abs/1501.04431) to justify our use of fractional counting for some of these indicators.

Measurement processes and Bibliometric data that are based on low numbers of publications are more likely to suffer from *small size effects*, where small (random) variations in the data might lead to very significant deviations and discrepancies. HEIs with only a few publications in WoS-indexed sources should therefore not be described by WoS-based indicators (alone). To prevent this from happening, threshold values were implemented. No Bibliometric scores will be computed for the institutional ranking if the institution produced less than 50 WoS-indexed publications during the years 2010–2013. This count is based on a full counting scheme where each publication is allocated in full to every main institution mentioned in the publication's author affiliate addresses. Where the institutional ranking relates to all research publication output, irrespective of the field of science, the three subject rankings relate to specific fields: computer sciences, medicine, and psychology (the delineation of the fields is explained earlier). The lower publication output

---

* For a brief explanation of the idea of core publications, see http://www.leidenranking.com /methodology/indicators#core-journals.

threshold for each field is set at 20 (full counted) WoS-indexed publications.* Both these cutoff points were approved in December 2013 by the UMR Advisory Board.

11.3.2.2.5 Innovative Scientometric Indicator: Interdisciplinarity

One of the indicators in UMR that we regard as a very innovative one is the inter-disciplinarity indicator. The inclusion of this indicator as part of the research indicators in UMR is driven by our wish to express an additional characteristic of an institution's research performance. When people want to assess the research performance of a university, they often look at Bibliometric indicators such as citation rates or the number of a university's research publications that belong to the top 10% most frequently cited publications worldwide. While these may be important indicators, we feel that there is more to research than just that. Important scientific discoveries these days often take place at the intersection of different disciplines. The frontiers of research are often at the edge of disciplines. And many of today's societal challenges are wicked problems that are multidisciplinary in character, requiring the use of methods and insights from several established disciplines or traditional fields of study. One can think of global warming. Studying it involves the study of ecological systems, human behavior, and energy science and political science to mention a few.

Therefore, the multidisciplinary of research reflects its innovative character. Using its large database of scientific publications per university, UMR looks at the reference list of the university's publications. All the publications in this list belong to one or more scientific field (if they belong to more than one, we distribute them using fractional counting). UMR is using the WoS scientific fields for this. There are many such fields, with some being very close to each other and others more far apart. In this way, the references of each publication may be distributed across a range of fields. It is good to know that the strength of the relationship between two fields is regarded as being large if the two fields are frequently making use of each other's literature. The degree of interdisciplinarity of a publication is minimal if all references belong to the same scientific field. The larger the spread of references across the various fields and the weaker the relationship between the fields, the bigger the degree of interdisciplinarity of a particular publication. Taking into account all publications of a university, UMR thus computes the degree of interdisciplinarity of a university's research publications.

Observing the interdisciplinarity scores for all institutions covered in UMR, we can conclude that our interdisciplinarity indicator is really quite different than any of the traditional Bibliometric indicators. Several well-known HEIs that score high on our citation rate indicator actually score relatively low on the interdisciplinarity

---

* For a detailed overview of indicators and methods used to calculate indicators, see http://www.umultirank.org/cms/wp-content/uploads/2016/03/Bibliometric-Analysis-in-U-Multirank-2016.pdf.

indicator. This result may be expected for cases of relatively specialized institutions. But some of the large comprehensive state universities also take a moderate position in terms of their interdisciplinarity score. A few medical colleges express quite a high degree of interdisciplinarity, and this also applies to some of the technical universities. We conclude from these findings that UMR reaches the parts that other rankings have not yet reached. It highlights dimensions of performance and research quality not included elsewhere.

11.4 Data Verification

Data verification and data checks are crucial elements of UMR. To ensure data quality to the highest level possible, UMR developed a set of procedures. A first element of data quality is to ensure comparability. To make sure that the data of the institution are comparable, the institutional and the subject questionnaires include guidelines and definitions of all the data items requested. The institutional and subject questionnaires include [?] symbols that provide information about the definitions that are being used and particular issues related to specific questions. Institutions participating in UMR can also contact UMR by e-mail if they have questions about the concepts used in the institutional and subject questionnaires (U-Multirank, 2015a, p. 9). Since our main focus is on the institutional level, we will explain further in the following the data verification of the institutional data collection. It has to be noted that the data verification follows a similar approach for the subject questionnaires.

11.4.1 Data Verification Institutional Questionnaire

The verification of the data provided in the institutional questionnaire consists of several steps. In the first phase, the participating institutions provide their data using the online questionnaire. These data are intensively checked by the UMR team. The data are checked by using both automated checks and manual checks for consistency, plausibility (including checks of outliers), and missing data.

The institutional questionnaire includes a number of automated checks. Once the institutions have provided the data for the institutional questionnaire and before they submit their institutional questionnaire to UMR, the institutions will see comments and questions on the data provided (if applicable) below the respective question concerned and on the final page. Institutions are asked to verify and correct the data of the questions before submitting the questionnaire. The automated checks will disappear once the data have been filled in correctly.

The manual checks applied to the institutional questionnaire concern checks on the preliminary scores of ranking and mapping indicators, checks on the consistency of the data, and checks on the consistency of the data compared to the data provided in a previous ranking of UMR (if applicable). If the UMR team has questions and/or comments on the data provided, then these are implemented in the institutional questionnaire and in this way communicated to the institutions.

Once the institutional questionnaires are checked, the second phase starts. Institutions are invited to respond to the questions and comments implemented by the UMR team in their institutional questionnaire. The institutions are to clarify, correct, and add data in their original questionnaire. After the final submission of the institutional questionnaires, the data are checked again in a similar way as in the first verification round. The difference with the first round is that the remaining questions and comments of the UMR team are communicated by e-mail to the institutions. Once all data submissions are finalized and the data are regarded as valid and complete, the indicator scores are calculated (U-Multirank, 2015a, p. 9).

11.5 Presentation of Data

11.5.1 Rank Group Methodology

The rank group calculation for the standard indicators is based on the median (per indicator) of the total sample.* Because of the fact that the scores on a number of indicators are not normally distributed, this standard method produces ranking groups in which the within-group variance is very large. Therefore, we applied a method which takes into account the distribution of scores in a better way. After testing and analyzing various methods, we decided to log normalize the scores for those indicators that had a skewed distribution and apply the standard grouping method to these log-normalized scores. To determine whether or not to use log-normalized scores, the ratio median/mean is calculated, and for all indicators that are outside the 25% bandwidth around 1 (– or + 12.5%), the log-normalized score is applied. This procedure was applied for the first time in the 2016 release of UMR both for institutional rankings and for the six new subject rankings. In order not to modify rank groups without changes in the underlying data in the 2014 and 2015 subject rankings, the new methodology will be applied only when updating these subjects in 2017.

UMR indicates how universities perform by showing their position in five performance groups (*very good* through to *weak*) for each of some 30 different indicators. For this, it uses five rank groups. The rank groups refer to the distance of the indicator score of an individual institution to the average—or rather the median—performance of all institutions that UMR has data for. With regard

---

* This procedure does not apply to the indicators based on the student survey and the two subject-based rating indicators on contact to work environment and international orientation of the programs. The rank group methodology is presented in detail in the web tool.

to the grouping procedure, there are three different types of indicators and rank group calculations:

1. *Regular* quantitative indicator
2. Rating indicators
3. Student survey indicators

We focus here on the first type.

Most indicators used are based on continuous measures on particular scales (e.g., the percentage out of a total; a relation A:B). For those indicators, the calculation of the five different groups is referring to the median (per indicator) of the total sample. UMR rank groups are defined in terms of distance of the score of an institution from the median (for a single indicator). Groups range from the best group A to the lowest group E:

1. Group A: If the value of the indicator is above the median plus 25% (value > median + 25%)
2. Group B: If the value of the indicator is less than or equal to the median plus 25% and greater than the median (median + 25% ≥ value > median)
3. Group C: If the value of the indicator is less than or equal to the median and greater than the median minus 25% (median ≥ value > median – 25%)
4. Group D: If the value of the indicator is less than or equal to the median minus 25% and above zero (median – 25% ≥ value > 0)
5. Group E: If the value of the indicator is zero (value = 0)

11.5.2 Criteria

A good classification method should largely reflect the pattern that underlies the data. In other words, any classification method that distorts the underlying pattern of data is considered a poor classification.

To produce results that are consistent, intuitive, and methodologically sound, there are three main criteria that will be used to test the alternative classification methods:

1. There should be HEIs in all four groups (the fifth group—value = 0—is treated in a different way).
2. The within-group variance should be minimized. That means that the HEIs in one group should be as much as alike possible.
3. Stability: The number of instances in which the alternative method would lead to a substantial change in group* has to be as limited as possible.

---

* If the group would go up or down two or three categories.

The standard UMR method is based on the median of the nonzero scores. The median is the cutoff point between categories 2 and 3. The cutoff points between categories 1 and 2 and between 3 and 4 are determined by a fixed percentage of the median: 25%.

The standard UMR method is a sound method for those indicators that have a normal distribution. For those indicators that have a skewed distribution,* the within-group variance in the top group is substantial.

For those indicators, two alternative methods are evaluated.

11.5.3 Alternative Grouping Methods

- The first alternative is to apply the standard method to log-transformed scores. The rationale is that the log-transformed scores have a normal distribution on which the standard UMR method can be applied.
- The second alternative is the head/tail breaks method. The head/tail breaks method is a classification method for data with a heavy tailed distribution. In this method, nested means are used to determine the cutoff points between the categories. It uses the average as a first cutoff point. The group above the average is divided in two by the average of that group, etc. Some details of this method can be found in the paper by Jiang (2013). If the distribution has a heavy tail (many cases with very low scores), the within-group variance can be reduced by applying this method. It uses the average as a first cutoff point.

11.5.4 Evaluation

Changing the method may have a benefit in terms of avoiding empty top groups and minimizing within-group variance (the cases within each group are more alike). The cost of changing method relates to the cost of implementation and the cost of loss of transparency/explaining the use of different methods and shifts in group scores.

The cost of using alternative 1 is relatively low. The standard method can be applied to all indicators, which is a good thing in terms of transparency. However, it has to be communicated clearly whether the categorization is based on the log-transformed data or the original data. The benefits in terms of minimizing within-group variance are rather limited.

The costs of combining alternative two are high, both in implementation and communication. The inconsistency of the results of this method and the results from the previous release are substantial. This leads to high costs in terms of communication and a high risk in terms of legitimacy and support. *People* will not understand why positions change while the underlying scores do not change. This

---

* Many indicators have a distribution that is skewed to the right: many low-scoring HEIs and a few high-scoring HEIs.

new method cannot be implemented in the upcoming release. The benefits are high as the within-group variance reduces significantly.

11.5.5 Conclusion

- The standard method will be applied to those indicators that have a normal distribution of scores.
- For those indicators where the scores are not normally distributed, the standard method is applied on the log-transformed scores.

11.6 UMR Web Tool

From the beginning of UMR, the aim has been to create transparency in a very diverse higher education area. To show this diversity in performance profiles and excellence, UMR developed a multidimensional approach. The results of the multidimensional *rankings* demonstrate that *the best* institution does not exist. What lists or overviews of HEIs emerge as the best performing depends on the preferences of the user and the indicators chosen. Different users may have different interests and may therefore come up with different lists of well-performing institutions (Federkeil et al., 2012a, pp. 168–172; U-Multirank, 2015a, pp. 13–14).

11.6.1 Development of the Web Tool

A print publication would not adequately represent this multidimensional, interactive, and user-driven ranking. UMR has therefore developed an innovative user-friendly web tool to present the results. The first step in the development of the web tool has been to define the basic functions and features. It has been important that the UMR web tool has to provide different user journeys to different users. In order to make the web tool user-oriented, the web tool offers a flexible user-driven ranking, the *Compare function* on the web tool, and a special track *For students* to take into account the information needs of students (U-Multirank, 2015a, p. 14).

11.6.1.1 Stakeholder Consultations

The web tool has developed in close consultation with the European Commission (Directorate-General for Education and Culture), the UMR Advisory Board, and other stakeholders, particularly students. A major challenge of developing the web tool has been to find a balance between the different needs of, on the one hand, the HEIs and, on the other hand, the users of the web tool. It is important that the data are presented in a way that conforms to how the HEIs prefer to see their performances presented while at the same time ensuring that the web tool is easy to use for its users. The stakeholders and user groups have been consulted on the conceptual approach of the web tool prototype to determine whether the functionality met the needs of the users and whether it addressed the concerns of the stakeholders regarding the representation of the data and the comparisons. These consultations have proven to be very effective (Federkeil et al., 2012a, pp. 168–172; U-Multirank, 2015a, pp. 10, 13–14).

11.6.2 Design and Distinctive Features of the Web Tool for Measuring Performance

The UMR web tool offers users four options to compare the performances of HEIs. These are the *For students* track, the *Compare* track, the *At a glance* track, and the *Readymade* track. Each of them is explained in more detail in the following.

11.6.2.1 User-Driven Personalized Rankings

A distinctive and major feature of the web tool is that it allows for personalized ranking tables. These personalized rankings reflect UMR's basic philosophy. An interactive user-driven approach has been implemented. Users have different opinions concerning the relevance of the indicators that should be included in a ranking. To take into account these different views, the web tool enables users to select their own individual indicators. This option is available for the institutional rankings as well as the subject rankings (Federkeil et al., 2012a, pp. 171–172).

Users are able to create personalized rankings through the *For students* and *Compare* tracks. The *For students* track has been designed particularly for student users and places emphasis on indicators related to teaching and learning. The *For students* track makes it possible for users to select institutional profiles or subject areas to narrow down their choice to only the HEIs that match with their preferences. The *Compare* track allows users to compare HEIs in three different ways. The first option is to compare similar universities on a like-with-like basis by defining the kinds of institutions that they are interested in. The second option is to compare HEIs by name. Users can select up to three institutions to compare with each other. The final option is to compare a particular HEI with other institutions that have a similar profile (U-Multirank, 2015a, p. 15).

The *For students* and *Compare* tracks allow users to create personalized ranking tables which can be sorted by the users in different ways. The HEIs can be categorized by alphabet, by individual indicators, or by top scores, where institutions with the highest number of A (very good) scores are shown first in the table. If there is more than one institution that has a similar number of A scores, then the number of B scores becomes relevant, and so forth. This interactive function of the web tool is a unique feature of UMR (U-Multirank, 2015a, p. 15).

11.6.2.2 At a Glance: Detailed Information on a Specific Institution

Next to the user-driven options to create personalized rankings, UMR also allows users to retrieve immediately the main results of a particular institution. This option has been introduced since not all the users will want to read a lengthy table when using UMR. The idea behind including this option is that it may encourage users to access data at a lower level to more detailed information (Federkeil et al., 2012a, p. 173).

This immediate route to the detailed result of a particular HEI can be found at the At a glance track. At this track, UMR presents the detailed profiles of HEIs from all over the world. The profiles provide the users with information on the type of institution and about the type of activities that they are good at. The information about the type of institution includes qualitative descriptive information, such as the address, country, and website of the institution. Regarding the performance on the activities, users are given detailed information on individual indicators concerning the score of the institution, the performance group, the minimum and maximum scores per indicator, and the distribution of the score across the full sample of institutions. In addition, the *At a glance* page contains information on the study programs included in the subject data (U-Multirank, 2015a, p. 15).

The main results of a HEI on the At a glance page are also visually represented. This graphic presentation provides insights into the institutional results with the performance of the institution as a whole. The institutional results are not aggregated into one composite indicator. The results are presented in a sunburst chart that gives an integrated graphical view of the performance of an institution. In short, the sunburst chart shows each of the five dimensions in a different color with the rays representing the individual indicators. The sunburst chart is an innovative way to present the performance and the profiles of HEIs (Federkeil et al., 2012a, p. 173; U-Multirank, 2015a, p. 15).

11.6.2.3 Readymade Rankings

UMR also developed readymade rankings, which are an example set of rankings. These readymade rankings have been developed to demonstrate to users the options and possibilities of the web tool. Several readymade rankings have been developed to show the users which HEIs are performing best in different areas. While it is not possible to produce a definitive list of the world's *top performing* universities in these areas, UMR shows top performances in different aspects of them (U-Multirank, 2015a, p. 15). At the moment, three readymade rankings are available online:

1. *Research and research linkages ranking*: This readymade ranking aims to show how selected UMR institutions are performing in terms of seven different

Bibliometric performance indicators in the areas of research and research linkages (U-Multirank, 2015b).

2. *Economic engagement ranking*: This readymade ranking aims to show how selected UMR institutions are performing in terms of 11 performance indicators in the areas of knowledge transfer and regional engagement (U-Multirank, 2016c).

3. *Teaching and learning rankings*: These readymade rankings aim to show how selected UMR institutions are performing in terms of performance indicators on teaching and learning for the subjects included in UMR. The number of performance indicators depends on the subject selected. For example, for the teaching and learning ranking on physics, we have eight performance indicators, whereas the total number of performance indicators for the subject field of psychology is seven (U-Multirank, 2015c,d).

11.7 Conclusion

In this chapter, we address several aspects of Data Analytics and Scientometrics that apply to UMR. We describe the dimensions and the indicators, how the data are collected and verified, and how the data are presented and are shared in an innovative user-friendly web tool. In Section 11.1, we give a brief overview of UMR by describing why UMR has been developed. In addition, we set out the five design principles that have been used to develop UMR. In Section 11.2, we address the dimensions and indicators covered in UMR. We provide an overview of the indicators covered in the UMR institutional questionnaire for each of the five dimensions. Moreover, we describe the subject ranking indicators, and we have described and presented the mapping and institutional descriptive indicators. We conclude this section by briefly reflecting upon the student survey. Section 11.3 focuses on the data sources that are used to collect the data. We describe the different data sources that are used to retrieve data from. UMR makes use of three different questionnaires that give institutions and students the opportunity to provide data. These include the institutional questionnaire, the subject questionnaires, and the student survey. In addition, data are retrieved from publicly available data sources, including national administrative data sources, national student surveys, and international data sources. Section 11.4 addresses how the UMR team verifies the data. We provide an example of how the data in the institutional questionnaire is being checked. In Section 11.5, we give an overview of how the data are presented. In this section, we demonstrate how we calculate the position of HEIs in the five performance groups and which criteria apply and which alternatives are possible. In Sections 11.6 and 11.7, we describe the development of the UMR web tool and we presented the distinctive features to measure the performance of HEIs.

UMR has evolved into a very rich and comprehensive tool for comparing HEIs and their performances. It seeks to serve a wide variety of users in their quest for comparable information on universities and their programs. This complexity is a major asset as it avoids the production of oversimplified list of institutions that can have detrimental effects on the diversity of national systems and the international higher education area and the prestige and public view on large parts of higher education (such as the institutions that are not on those lists). But the thickness of its conceptual fabric and the complexity of possible views also pose a major challenge. Users have to invest in the tool and explore the potential that lies within to answer their specific questions. UMR has difficulty producing the sound bites that are needed to draw the attention that is needed to create the impact that lies in it. In addition to the development of new indicators and constant review of the existing ones, UMR will reach out to other databases and tools in the field to explore the options for a common (data) language that will allow the user to navigate on the ever-swelling sea of information. Interaction with other data providers and with the various users is key in meeting the challenges ahead.

References

Callaert, J., Epping, E., Federkeil, G., Jongbloed, B., Kaiser, F., & Tijssen, R. (2012). Data collection. In F.A. van Vught & F. Ziegele (Eds.), *Multidimensional Ranking: The Design and Development of U-Multirank* (pp. 125–134). Springer, the Netherlands.

European Tertiary Education Register. (n.d.). *ETER in a Nutshell*. Available online at https://www.eter-project.com/about/eter.

Federkeil, G., File, J., Kaiser, F., Van Vught, F.A., & Ziegele, F. (2012a). An interactive multidimensional ranking web tool. In F.A. van Vught & F. Ziegele (Eds.), *Multidimensional Ranking: The Design and Development of U-Multirank* (pp. 167–178). Springer, the Netherlands.

Federkeil, G., Jongbloed, B., Kaiser, F., & Westerheijden, D.F. (2012b). Dimensions and indicators. In F.A. van Vught & F. Ziegele (Eds.), *Multidimensional Ranking: The Design and Development of U-Multirank* (pp. 97–124). Springer, the Netherlands.

Federkeil, G., Kaiser, F., Van Vught, F.A., & Westerheijden, D.F. (2012c). Background and design. In F.A. van Vught & F. Ziegele (Eds.), *Multidimensional Ranking: The Design and Development of U-Multirank* (pp. 85–96). Springer, the Netherlands.

Federkeil, G., Van Vught, F.A., & Westerheijden, D.F. (2012d). A evaluation and critique of current rankings. In F.A. van Vught & F. Ziegele (Eds.), *Multidimensional Ranking: The Design and Development of U-Multirank* (pp. 39–70). Springer, the Netherlands.

Federkeil, G., Van Vught, F.A., & Westerheijden, D.F. (2012e). Classifications and ranking. In F.A. van Vught & F. Ziegele (Eds.), *Multidimensional Ranking: The Design and Development of U-Multirank* (pp. 25–37). Springer, the Netherlands.

Higher Education Statistical Agency (n.d.). What we do and who we work with. Available online at https://www.hesa.ac.uk/about/what-we-do.

International Association of Universities (2016). World Higher Education Database. Available online at http://www.whed.net/About.html.

Jiang, B. (2013). Head/tail breaks: A new classification scheme for data with a heavy-tailed distribution. *The Professional Geographer*, 65(3):482–494.

National Center for Education Statistics (n.d.). *Integrated Postsecondary Education Data System, About IPEDS*. Available online at https://nces.ed.gov/ipeds/Home/AboutIPEDS.

U-Multirank (n.d.). *Data sources and data verification*. Available online at http://www.umultirank.org/#!/about/methodology/data-sources?trackType=about&sightMode=undefined&section=undefined.

U-Multirank. (2015a). *Implementation of a user-driven, multi-dimensional and international ranking of HEIs: Final report on project phase 1*. Available online at http://publications.europa.eu/en/publication-detail/-/publication/95b37c1f-cb09-11e5-a4b5-01aa75ed71a1/language-en.

U-Multirank (2015b). *Ranking description research and research linkages ranking*. Available online at http://www.multirank.eu/fileadmin/downloads/readymades/Rankings_Description_Institutional_Research_Research_Linkages_UMR_2015.pdf.

U-Multirank (2015c). *Rankings description teaching and learning physics*. Available online at http://www.multirank.eu/fileadmin/downloads/readymades/Rankings_Description_Fieldbase_Teaching_Learning_Physics_UMR_2015.pdf.

U-Multirank (2015d). *Rankings description teaching and learning psychology programmes*. Available online at http://www.multirank.eu/fileadmin/downloads/readymades/Rankings_Description_Fieldbase_Teaching_Learning_Psychology_UMR_2015.pdf.

U-Multirank (2016a). *Indicator Book 2016*. Available online at http://umultirank.org/#!/about/methodology/data-sources?trackType=about&sightMode=undefined&section=undefined.

U-Multirank (2016b). *Ranking description economic engagement ranking*. Available online at http://www.umultirank.org/cms/wp-content/uploads/2016/12/Scroll-over-description_Economic-Engagement_RMR-2016.pdf.

U-Multirank (2016c). *U-Multirank Newsletter Issue IV*. Available online at http://www.umultirank.org/cms/wp-content/uploads/2016/04/January-2016-Newsletter.pdf.

U-Multirank—International ranking of universities and university colleges. (2013). HiOA. N.p., Nov. 18, 2013. Web. Nov. 24, 2016. Available online at https://www.hioa.no/eng/About-HiOA/System-for-kvalitetssikring-og-kvalitetsutvikling-av-utdanningene-ved-HiOA/U-Multirank-International-ranking-of-universities-and-university-colleges.

Van Vught, F., & Ziegele, F. (2011). Design and testing the feasibility of a multidimensional global university ranking. Final Report. European Community, Europe: Consortium for Higher Education and Research Performance Assessment, CHERPA-Network.

Van Vught, F.A., & Ziegele, F. (2012). Concluding remarks. In Vught & F. Ziegele (Eds.), *Multidimensional Ranking: The Design and Development of U-Multirank* (pp. 179–190). Springer, the Netherlands.

Waltman, L., & Van Eck, N.J. (2015). Field-normalized citation impact indicators and the choice of an appropriate counting method. *Journal of Informetrics*, 9(4):872–894.

Chapter 12

Quantitative Analysis of *U.S. News & World Report* University Rankings

James Fangmeyer Jr. and Nathalíe Galeano

Tecnológico de Monterrey, Monterrey, Mexico

Contents

This is the story of a first mover and how it continues to move higher education. *U.S. News & World Report* (U.S. News) first published *America's Best Colleges*, which ranked colleges and universities in the United States, in 1983. This was the first major higher education ranking in the world, and it became extremely popular. Thirty years later, U.S. News was not, however, the first organization to produce an international ranking comparing universities across different countries. U.S. News finally entered the world ranking stage with its inaugural *Best Global Universities* (BGU) in 2014. Globalization, advances in Data Analytics and Scientometrics, and rising demand for higher education have produced a proliferation of rankings and entrenched their position in society. For better or for worse, millions of educators, students, and parents consult university rankings each year. The rankings influence decisions that control millions of dollars and brilliant ideas, and *U.S. News & World Report* continues to be a major actor in this environment.

This chapter reviews the impact of U.S. News university rankings in science and culture. Section 12.1 recounts the magazine's ranking history and recent activity. Section 12.2 explains methodology from the BGU ranking. The scientific relevance of U.S. News university rankings is developed by means of a literature review in Section 12.3. The first part of Section 12.3 is a Scientometric review of 233 documents related to U.S. News rankings. Questions asked include, Who publishes about U.S. News? Which academic disciplines research U.S. News? Where is research on U.S. News published? The second part of Section 12.3 organizes the contents of the articles to understand perspectives and preoccupations around the rankings. The story would not be complete without a supporting cast; U.S. News is compared with other international rankings in Section 12.4. Finally, this chapter concludes in Section 12.5 by predicting future directions for global university rankings and U.S. News in particular.

12.1 *U.S. News & World Report* and America's Best Colleges Ranking

Before its first university ranking, *U.S. News & World Report* was a weekly magazine that circulated stories about economy, health, and education rather than sports, entertainment, and celebrities. Its primary competitors for readership were *Time* and *Newsweek*. It had begun as two separate magazines, both founded by journalist David Lawrence and run out of Washington, DC. The magazines were merged into *U.S. News & World Report* in 1948 (Sumner, 2012). To give a sense of scale, the magazine passed the threshold of 2 million monthly readers in 1973.

The first edition of America's Best Colleges was published by *U.S. News & World Report* in 1983. The first 1983 ranking was based on survey responses of several hundred university presidents, who were asked to opine on the relative quality of America's higher education institutions (Leiby, 2014): Stanford came in first, Harvard second, next Yale, and then Princeton. One year after publishing the first America's Best Colleges ranking, *U.S. News & World Report* was purchased by the owner of *New York Daily News* (*U.S. News & World Report*, 2013). Rankings were not published in 1984, but have been published every year since, proving to be a readership magnet.

America's Best Colleges has been published as a separate guidebook since 1987, and it is exceptionally popular online. On the release of the 2014 rankings, 2.6 million unique visitors generated 18.9 million page views in 1 day at USNews.com (Smith, 2013). The website has more than 20 million unique visitors each month, 10 times the readership that *U.S. News & World Report* magazine had in 1973.

Great popularity often comes with at least a measure of detraction. U.S. News ranking methodology has received public criticism since at least 1995, when Reed College refused to submit the reputation survey that had been part of the ranking since 1983. The reputation survey has been one of the most contentious parts of America's Best Colleges methodology for its subjective nature and monolithic outcome. Other points of contention regard the use of self-reported indicators, the close correlation between a university's wealth and rank, and the lack of indicators which accurately capture the quality of education at each school.

12.2 *U.S. News & World Report* and the *BGU* Ranking

In October 2014, *U.S. News & World Report* launched a new ranking with a new methodology: the 2015 BGU ranking. In 2014, the ranking covered 500 institutions from 49 countries across the globe and evaluated them based on a methodology that included, among other new indicators, Bibliometric evidence of a university's performance based on publications and citations. The top 3 schools were Harvard, Massachusetts Institute of Technology, and University of California–Berkeley. Of the top 500 schools, 134 featured from the United States, 42 from Germany, 38 from the United Kingdom, and 27 from China. Rob Morse of U.S. News, said, "I think it's natural for U.S. News to get into this space. . . . We're well-known in the field for doing academic rankings so we thought it was a natural extension of the other rankings that we're doing" (Redden, 2014). Although it may be a natural extension, the methodology applied in *U.S. News & World Report* Best Global Universities differs greatly from the methodology for America's Best Colleges (Morse et al., 2016).

The first major distinction between America's Best Colleges and the BGU international ranking is that BGU uses Bibliometric data, a type of Scientometrics not used in the national rankings. Bibliometric data used in the BGU ranking include publications, citations, and coauthorship. The publications considered are those from a recent 5-year period, and the citations considered are those attributed to the publications from this 5-year period, even if they come from publications outside of the 5-year period. For example, the 2017 ranking uses publications from 2010 to 2014 and citations up to April 2016 attributed back to the documents from 2010 to 2014. Data for the 2015 and 2016 rankings were provided by Thomson Reuters IP & Science via its Web of Science product, which tracks around 13,000 scientific publication sources including academic journals, conferences, and books. Thomson Reuters IP & Science was purchased in 2016 by the Onex Corporation and Baring Private Equity Asia and spun into a new company called Clarivate Analytics. Data for the 2017 BGU were provided by Clarivate Analytics.

The second major change is in the eligibility procedure. U.S. News establishes a list of eligible universities from around the world with a two-step process. The first step is the Clarivate Analytics Academic Reputation Survey. This is an invitation-only survey sent to authors selected from the Web of Science database. The survey is available in 10 languages. Respondents are asked to evaluate programs in the disciplines with which they are familiar. This creates results at the department level rather than the institutional level and gives a more complete picture of the university. Unique responses in 2017 totaled around 50,000, which were weighted based on geographic region to correct for differing response rates. Finally, scores from surveys over the past 5 years are aggregated to produce the score for the ranking year. For 2017, the top 200 universities in the Clarivate Analytics Academic Reputation Survey were added to the list.

Next, the institutions with most scholarly publications in a recent 5-year period are added to the list. For the 2017 ranking, the 1,000 institutions that had the most publications from 2010 to 2014 were considered eligible for ranking. The list of 1,000 institutions was de-duplicated for schools already in the top 200 of global reputation, and because some schools tied with the same number of publications, the list is not exactly 1,000 items. Thus, the final list of eligible universities for 2017 included 1,262 institutions.

Finally, BGU uses 12 weighted indicators to rank the schools in the ranking universe. Two indicators are based on reputation surveys, and the other 10 are based on Bibliometric data, as follows:

1. Global research reputation (12.5%): An aggregation of 5 years of scores in the Clarivate Analytics Academic Reputation Survey. The scoring scale of the survey and the aggregation weights over the five previous years are not described on the website USNews.com. The top 200 schools in this indicator are automatically considered eligible for the ranking, regardless of quantity of publications.
2. Regional research reputation (12.5%): Respondents are asked to rate universities in their region. The score is the aggregate of 5 years of results. Regions are determined by the United Nations definition.
3. Publications (10%): Total number of scholarly articles, reviews, and notes published in a recent 5-year period and indexed in the Clarivate Analytics Web of Science database. The top 1,000 schools, de-duplicating for the top 200 reputation schools, are eligible for ranking, regardless of what reputation they may have.
4. Books (2.5%): The number of books indexed in the Web of Science.
5. Conferences (2.5%): The number of publications in conference proceedings that are indexed in the Web of Science.
6. Normalized citation impact (10%): This is the number of citations per paper, but normalized to control for several forces. It is normalized by publication year, publication type, and research discipline, such that an average publication across these factors has a score of 1, and a score above 1 reflects that the publication in question received more citations than its peers.

7. Total citations (7.5%): Calculated by multiplying the publication indicator by the normalized citation impact indicator. Thus, this is not a simple count, but rather a normalized indicator.

8. Number of publications that are among the 10% most cited (12.5%): Publications are compared to publications of the same research area, document type, and year.

9. Percentage of total publications that are among the 10% most cited (10%): The percentage of a university's total papers that are in the top 10% most cited papers with the same research area, document type, and publication year.

10. International collaboration (10%): Proportion of a university's papers with an international coauthor, controlling for the university's country. Control is done by dividing the university's proportion of papers with an international coauthor by the proportion of all papers from the university's country with an international coauthor.

11. Number of highly cited papers that are among the top 1% most cited in their respective field (5%): Publications in the top 1% by subject area and by year. Documents from the last 10 years are eligible in 22 different subject areas in the Web of Science.

12. Percentage of total publications that are among the top 1% most highly cited papers (5%): Number of highly cited papers divided by the total number of papers produced by a university.

After retrieving the indicator values, these are transformed from indicator values into z-scores, which measure values in terms of standard deviations and make comparison between different types of data more robust. Log transformation is used before calculating z-scores for these highly skewed variables: publications, books, conferences, total citations, number of publications that are among the 10% most cited, global research reputation, regional research reputation, number of highly cited papers that are among the top 1% most cited in their respective field, and international collaboration. U.S. News reports that there were no missing Bibliometric or reputation indicators for the BGU 2017; however, citation data from arts and humanities journals were intentionally omitted. U.S. News reports that arts and humanities journals do not receive many citations, and therefore, omitting this information makes the results more robust. Publication counts from these journals were included in publication indicators.

The minimum score from the ranking universe is subtracted from the score of all universities, leaving the lowest-scoring school with a score of zero. Then, the highest performing school is rescaled to have a score of 100, and each school is rescaled proportionally to the highest score. Overall scores are rounded to one decimal place. The final 2017 ranking includes the highest scoring 1,000 schools from 65 countries, up from 750 schools from 57 countries in 2016 and from 500 schools in 49 countries in 2015.

12.3 Scientometric Review

U.S. News university rankings and their impact have been a topic of academic study for over 20 years. A review of academic literature in the Elsevier Scopus database was performed to understand the scientific perspective on U.S. News university rankings. The Elsevier Scopus database tracks documents published in approximately 20,000 journals and conferences worldwide. The review was done by searching in the title, abstract, and keywords of documents for *U.S. News* or *US News* along with the root word *rank*. The exact query that was run on September 29, 2016, in the Scopus Advanced Search feature was (TITLE-ABS-KEY ("U.S. NEWS") OR TITLE-ABS-KEY ("US NEWS")) AND TITLE-ABS-KEY (RANK*). As indicated by the quotation marks around the search terms, the exact and complete phrases *US News* and *U.S. News* were searched to eliminate unrelated results that contained only *US, U.S.,* or *News*. The search is not case sensitive. The asterisk in the query after the root word *rank* allows results containing *ranking* or *ranked*. Since 1992, 233 publications match the query above. The new ranking is still unexplored in peer-reviewed research.

Each of the 233 publications returned in the survey was manually reviewed to assure relevance. Seven documents were discarded because they had no relevance to U.S. News rankings. Another 50 documents were discarded because they were relevant to U.S. News best hospital rankings, but not university rankings. One document was discarded as a duplicate record. The remaining 175 documents date from 1994. Just two of these publications contain the term *Best Global Universities*: one published in 2014 and another in 2016. These two articles show potential interest for BGU in the academic community, but there is little documentation on the ranking's performance and impact.

The first finding is that the rate of research on U.S. News university rankings is increasing over time (see Figure 12.1 for complete history). On average over 23 years, eight publications/year are related to U.S. News university rankings. In the last 10 years, an average of 13 publications/year is related to U.S. News rankings. Academic coverage of U.S. News rankings increased markedly around 2000, at the start of the Internet era.

Almost two thirds of the scientific reporting on U.S. News is done through journal articles. About one-fifth is done in conference papers, and the remaining corpus is composed of review papers, books, and notes (Figure 12.2). The second finding is that research on U.S. News is almost twice as likely as a random Scopus publication to be published in conference papers. Twenty percent of research on U.S. News university rankings is published in conference papers, while 11.8% of all Scopus-indexed publications are found in conference papers (Elsevier, 2016). This has to do with publication tendencies by discipline.

Figure 12.3 shows the disciplines that most contribute to research on U.S. News university rankings. The percentages sum to more than 100% because one publication can be listed in more than one discipline. Social science is the field most

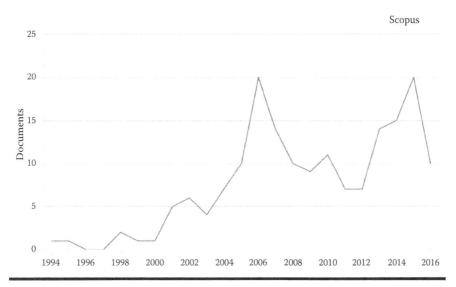

Figure 12.1 Scopus-indexed publications related to U.S. News university rankings. (From Elsevier, http://www.scopus.com, Amsterdam. Copyright © 2016 Elsevier B. V. All rights reserved. Scopus® is a registered trademark of Elsevier B. V.)

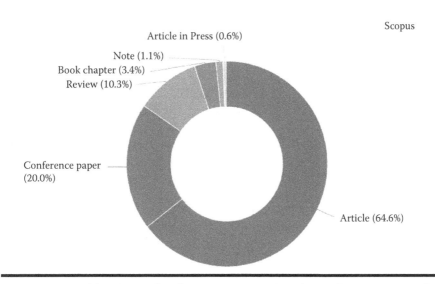

Figure 12.2 Publications related to U.S. News university rankings in Scopus by document type. (From Elsevier, http://www.scopus.com, Amsterdam. Copyright © 2016 Elsevier B. V. All rights reserved. Scopus® is a registered trademark of Elsevier B. V.)

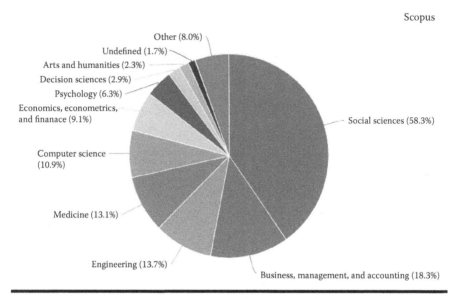

Figure 12.3 Distribution of Scopus disciplines in which U.S. News studies are published. Percentages do not sum to 100% because one article can be classified in more than one discipline. (From Elsevier, http://www.scopus.com, Amsterdam. Copyright © 2016 Elsevier B. V. All rights reserved. Scopus® is a registered trademark of Elsevier B. V.)

concerned with U.S. News rankings, involved in more than half of all publications on the topic. The third major finding is that education research is the key to understanding publishing tendencies on U.S. News university rankings. Looking at subcategories of social science, it is found that the education discipline is responsible for producing much of the research on U.S. News in conference papers. Education is a subcategory of social science that produces 24% of its publications in conference papers (Elsevier, 2016), and the source that has published the largest number of documents on U.S. News university rankings is the *American Society for Engineering Education* (ASEE) *Annual Conference and Exposition, Conference Proceedings* (marked by an asterisk in Figure 12.4).

Nine Scopus sources have published three or more articles related to U.S. News university rankings since 1994, seen in Figure 12.4, and seven of these nine sources are journals focused on education. Three of these specifically focus on higher education and two on engineering education. Law, medicine, social work, and marketing each feature one time as journal topics. There are 15 Scopus sources with two publications about U.S. News since 1994 and 95 sources with one publication about U.S. News in that time. In total, 120 Scopus sources have published 175 documents about U.S. News university rankings in the last 23 years.

U.S. News university rankings can be further understood through data at the country level, institutional level, and author level. The fourth finding is that

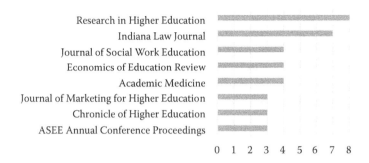

Figure 12.4 Scopus sources with the most publications related to U.S. News rankings.

research on U.S. News university rankings is highly concentrated by country. As determined by the address of the author's affiliation, the United States is the primary contributor of research on U.S. News university rankings. The United States produces 85% of the research in this review (148 documents). Scopus tracked a total of 13.7 million scientific documents between 2011 and 2015. Of these, the United States was responsible for 3.1 million, or 23%. The United States produces just 23% of all the world's research, but 85% of research on U.S. News university rankings. Despite the skew to the United States, Europe, Asia, and Africa also produce research on U.S. News university rankings. China is the second largest producer of research on U.S. News university rankings with eight documents, and the United Kingdom is third with three documents.

The fifth major finding is that the distribution of institutions publishing on U.S. News university rankings is flat and long; no school has more than six publications since 1994, and there are 160 institutions with at least one publication. The average number of publications for each institution in the survey since is 1.5, the median is 1, and the mode is 1. The institutions with the most publications in the survey tend to be large institutions with graduate schools and broad research output. The seven schools producing the most research on U.S. News university rankings are five state universities (University of Georgia, Pennsylvania State, Indiana University, University of Kansas–Lawrence, and Michigan State) and two Ivy League private schools (Cornell and Columbia). Pennsylvania State is also among the seven schools that produce the most education research in the world.

The sixth finding is that the distribution of authors is also flat and long and perhaps that scholar-administrators have more interest in the rankings than nonadministrators. There are 157 authors in total in the surveyed literature. Production is remarkably nonconcentrated: 80% of the publications are attributable to the top 75 percentiles of the author population. The modal author has contributed one paper on the topic; 51 authors, one-third of the total, have produced two or more papers on the subject. The most represented author is Michael J. Holosko of the

University of Georgia School of Social Work, who has six documents in the survey. Small collaborative teams are common. The average paper on the topic has 2.4 authors, and 70% of papers have at least two authors. The maximum number of authors on one publication is nine. There is some evidence to support the idea that research on rankings is mainly done by scholar-administrators. Two of the six authors with the most publications in the survey are deans: Andrew P. Morriss is dean of Texas A&M University School of Law, and Michael C. Roberts is dean of Graduate Studies at Kansas University.

The seventh finding is that 1,106 Scopus-indexed documents have cited the 175 publications in this survey (including publications in the survey citing other publications in the survey). The first of these referencing documents was published in 2000, 6 years after the first research published on U.S. News university rankings. Citing documents are more likely to be articles (74%) than the documents that they are citing (recall from Figure 12.2 that 65% of the reviewed literature comes from journal articles). The citing documents represent the same subject disciplines as the cited documents, except for engineering, which is much less likely to cite work on U.S. News university rankings than it is to produce such work. *Research in Higher Education* and *Scientometrics* are the journals most likely to cite work about U.S. News university rankings; 19 articles from each of these journals have cited publications found in this survey.

What do the 175 reviewed articles say about U.S. News rankings? For one, they say that U.S. News university rankings have become a substantial topic of research. More specifically, index keywords are one tool available in Scopus to compare the contents of these documents. Index keywords are a set of keywords created by Scopus that are systematically assigned to documents based on their contents. Additionally, there is no limit to the number of index keywords that can be tagged to a document, making index keywords likely to repeat across documents and identify similar documents.

In the 175 surveyed documents, the most common index keyword is *United States*, with 51 appearances, showing the preponderant connection that U.S. News rankings have to U.S. schools and audiences. Judging by index keywords, the two most researched fields are medicine and engineering. The root term *medic* appears 82 times in the dataset. The root term *engineer* appears 60 times. A small number of studies concern law schools or business schools. Revealing the nature of rankings as a data analysis tool, *data* and *statistic* appear 22 and 13 times in the index keyword results.

Especially captivating to researchers are the psychological and sociological impacts of U.S. News university rankings. U.S. News has caused a rue for condensing the social psychological perception of universities to the numbers that represent their ranks. Many studies in the review have intellectual heritage in quantitative authority (Porter, 1995) and commensuration theory (Espeland & Stevens, 1998). Scientific research found that U.S. News university rankings changed the traditional ways that status was created and maintained among U.S. schools (Sauder, 2006).

U.S. News university rankings also affect application decisions made my students and admission decisions made by schools (Sauder & Lancaster, 2006). These effects are especially felt at the top of the ranking, where breaking into the top 50 and relatively improving within the top 25 correlate with more first-year applications in the following cycle (Bowman & Bastedo, 2009). Not only applications, but also socioeconomic and racial demographics of top universities may be impacted by changes in rank (Meredith, 2004). The ranking's impact is not limitless, however. Research found that ranking outcomes have little correlation with admission outcomes at historically black colleges and universities (Jones, 2016).

Scientometric research frequently relies on U.S. News as a way to establish a dataset of the best authors or schools (Forbes et al., 2016; Holosko et al., 2016). Whether the authority of U.S. News is contested or taken for granted, it cannot be denied that these rankings influence how the discussion on higher education is framed.

This is the impact of U.S. News university rankings: a research topic that did not exist 25 years ago now appears in almost 200 publications, and these have been cited in over 1,100 publications. The rankings are of interest to researchers in several disciplines from the social to the hard sciences. U.S. News university rankings are a common topic at conferences, where representatives from many universities meet, evaluate, and compare themselves to one another. The rankings are a shared phenomenon; by organizing institutions regardless of geography, mission, or history, the rankings make it harder for universities to exist in isolation from one another.

12.4 U.S. News Rankings Compared and Discussed

One recent study found that 49 institutions appear in the top 100 in each of the six major rankings (Shehatta & Mahmood, 2016). The study found even greater correlation between these six rankings within the top 200 schools. Despite somewhat similar outcomes, the rankings have important differences in terms of data, methodology, and branding, which are discussed here. The six main Scientometric university rankers are U.S. News, Times Higher Education (THE), QS, the Academic Ranking of World Universities (ARWU), U-Multirank, and the National Taiwan University Ranking (NTU).

Choice of data is an investment at an international ranking agency because it takes time to establish relationships with data providers and create efficient data integration processes. The main sources of global Bibliometric data are the Scopus database produced by Elsevier and the WoS database, which was traditionally produced by Thomson Reuters but was purchased in October 2016 by the Onex Corporation, of Canada, and Baring Private Equity Asia and transferred to a new company called Clarivate Analytics. Both of these databases track quality publication sources including academic journals, conferences, and books, among others, and report all of the documents published by these sources. The exact selection of sources varies between the two databases and therefore changes the composition of the resulting datasets. Scopus indexes around 20,000 sources while WoS indexes around 13,000. U.S. News BGU ranking currently uses data collected in WoS, as does the ARWU, UMR, and the NTU ranking. THE used WoS until the end of 2014, when it switched its data provider to Scopus (THE, 2014). QS also uses data from Scopus.

Rankers have a range of possibilities in setting their methodologies. Different indicators can be included, and different weighting schemes can be applied to the indicators. U.S. News BGU uses two reputation indicators and 10 Bibliometric indicators. THE uses 13 indicators broken into five categories: teaching, research, citations, international outlook, and industry income; the first two categories (teaching and research) each have one reputation-based indicator. ARWU uses just six indicators, none of them based on reputation surveys. UMR uses over 100 indicators which can be combined in numerous ways to compare universities that are similar in selected dimensions; some of these indicators are based on surveys of students. NTU uses eight Bibliometric indicators and none based on reputation surveys.

Indicators unique to U.S. News BGU include number and percentage of a university's publications in the top 1% of the most cited publications. These give an intentional and large boost to a very elite set of research. Additionally, U.S. News BGU is the only ranking to use a regional research reputation indicator. In this indicator, respondents can only opine on programs in their same geographic region, per United Nations geographic definitions. The effect of this mandatory regional filter is to increase international diversity of the final results.

As a result of different branding, each ranking reaches different audiences. Scopus-indexed documents written on the various rankings reveal how widely each ranking has reached the scientific audience. U.S. News has by far the greatest measurable impact in the scientific community. Table 12.1 shows Scopus queries for major rankings and their returned relevant documents. U.S. News university rankings are relevant in 2.5 times more documents than their major ranking peers. U.S. News is relevant in 175 Scopus documents, while the ARWU, THE rankings, and QS rankings are each relevant in around 70 documents since 1994. Minor players UMR and the NTU rank both make around five appearances in scientific literature. Often, several rankings feature together in the same document.

Geography is another part of branding. Despite the fact that universities from the U.S. occupy most top positions in all university rankings, most rankings are not produced in the United States. U.S. News is the only Scientometric global university ranking produced in the United States, and its geography is inherent in its brand name. THE and QS are produced in England. The ARWU is produced in Shanghai, China, and is sometimes called the Shanghai Ranking, synonymous with its geographical origin. UMR is produced as an initiative of the European Commission with leading partners in Germany and the Netherlands, and the NTU rank is obviously produced in Taiwan.

Table 12.1   University Rankings Represented in Scientific Literature

Scopus Advanced Search Query	*Relevant Documents*
`(TITLE-ABS-KEY ("U.S. NEWS") OR TITLE-ABS-KEY ( "US NEWS")) AND TITLE-ABS-KEY (RANK*))`	175
`(TITLE-ABS-KEY ("Academic Ranking of World Universities")OR TITLE-ABS-KEY ("ARWU"))`	75
`(TITLE-ABS-KEY ("Times Higher Education") AND TITLE-ABS-KEY (rank*))`	73
`(TITLE-ABS-KEY ("QS") OR TITLE-ABS-KEY ( "Quacquarelli Symonds") AND TITLE-ABS-KEY (rank*))`	71
`TITLE-ABS-KEY ("U-Multirank")`	6
`(TITLE-ABS-KEY ("National Taiwan University") OR TITLE- ABS-KEY ("NTU") AND TITLE-ABS-KEY (rank*))`	4

Arbitrary differences in data, methodology, and branding make each ranking slightly different. This adds richness to the comparison of rankings. It also should ideally capture the wide diversity inherent in the universities that are being ranked. Hopefully, only the rankings that can offer the most useful perspective on global higher education will continue to appear in scientific literature and the public consciousness.

12.5 Future Directions for U.S. News and University Rankings

Major forces changing rankings come from readers, competitors, and technology. U.S. News launched BGU in 2015, signaling new priorities in international rankings and Bibliometric data analysis. U.S. News has updated the BGU methodology each year from 2015 to 2017, giving further clues to its future direction. For 2017, U.S. News dropped two indicators related to teaching from the methodology and added two indicators related to citations. The two dropped indicators had measured the number of PhDs awarded and the number of PhDs awarded per academic staff member. In their place, the number of highly cited papers (the top 1% in their field) and proportion of all papers that are highly cited were added. This changes the way that the real world is represented to the readers by shifting the ranking's focus. It is an example of U.S. News's pivot toward Bibliometric data.

Readers are increasingly moving from printed media to digital online media. The trend to online material has the potential to widen the geographical base of readers. Although it is not public information, many of U.S. News's 20 million monthly online readers must come from outside of the United States. U.S. News is better positioned to serve these readers by including their local institutions in the BGU ranking. As international travel and communication become easier, U.S. News is also serving a U.S. audience that wants to attend foreign universities. U.S. News has expanded the scope of BGU, moving from 500 ranked universities in 2015 to 1,000 in 2017. All of this has the potential to expand U.S. News's relevance to more readers.

University rankers compete for readers, event attendance, and advertising dollars. To differentiate itself, U.S. News states in its 2017 BGU methodology that it is the only ranker to use a regional research reputation indicator, which has "the effect of significantly increasing the international diversity of the rankings, since it focused on measuring academics' opinions of other universities within their region" (Morse et al., 2016). Other ranking agencies, however, have made regional presence felt by hosting regional summits attended by university directors. THE and QS both host summits around the globe that focus on regional issues, topics in higher education, and their own rankings. These events are marketed to university leaders. Rankers compete for the favor and trust of university administrators, who are looking for accurate and complete measurements of their own performance. These allegiances shake out in advertising dollars. Ranker websites and printed materials are covered with advertisements for universities from around the globe, and universities spend large sums of money to host THE and QS events. Competition may persuade U.S. News to host summits like some of its peers.

Add technological change to the mix. Data are cheaper and easier than ever to store and analyze. This is the major force behind the rise of Scientometric rankings. The world wide web was not a practical tool in the 1980s when U.S. News began ranking universities; it thus mailed surveys to individual influential leaders. Today, U.S. News pulls data on millions of published documents from the Internet and can easily calculate sums and averages across schools, subject areas, and geographical regions. BGU is almost entirely based on publication data, with just two reputational survey components. America's Best Colleges ranking of schools within the United States, however, is still full of self-reported indicators collected from each school, such as Alumni Giving Rate, and performance data available from the government or other bodies, such as Faculty Resources. The future direction for America's Best Colleges could be the inclusion of more Bibliometrics. The data supplier for U.S. News rankings, Thomson Reuters IP & Science, was recently purchased and spun off into Clarivate Analytics. Changes to the data products or simple uncertainty could potentially provoke U.S. News to change its data source.

Technology is also changing higher education through massive open online courses (MOOCs), and rankings have yet to take notice. A university's online educational offering is part of that institution's overall brand, impact, and expertise. No major ranking yet includes indicators such as number of MOOCs offered, students enrolled in MOOCs, or rating of MOOCs by users. Scientometrics currently used in global university rankings mostly measure research because publication and citation data are widely available and comparable. MOOCs inherently generate different, and probably more, data than a traditional physical classroom. All of this could create the next wave of Scientometrics around teaching. Given that major MOOC platforms such as Coursera, EdX, Udacity, and others are from the United States, U.S. News may have a local advantage to first collect data from these organizations and incorporate MOOC indicators into rankings.

Many future directions are possible; some of the most feasible based on the history and current position of U.S. News were presented in this section. The first university ranker has to prove itself again on the international stage, where rankers from Europe and Asia have more experience. The increased presence of Data Analytics in the world and the increased presence of Bibliometrics within U.S. News BGU rankings are trends to watch. Launching a global ranking and incorporating publication and citation data are U.S. News's most significant pivots in the last 3 years. U.S. News has twice the representation of other rankings in scientific literature. U.S. News university rankings are often studied in education research in conference papers. They are a main or relevant topic in 175 Scopus documents since 1994, and these documents have been cited in 1,106 Scopus documents. Scientific investigation has shown that U.S. News rankings impact applications, admissions, and status. It remains to be seen if this influence grows or diminishes over time. Ranking agencies such as U.S. News have influenced higher education for over 30 years, and there should be new developments aimed to continue their impact in the coming years.

References

Bowman N.A. & Bastedo M.N. (2009). Getting on the front page: Organizational reputation, status signals, and the impact of U.S. News and World Report on student decisions. *Research in Higher Education*, 50, 415–436.

Elsevier (2016). *Scopus Content Coverage Guide*, Elsevier, Amsterdam, https://www.elsevier .com/__data/assets/pdf_file/0007/69451/scopus_content_coverage_guide.pdf.

Espeland W.N. & Stevens M. (1998). Commensuration as a social process. *Annual Review of Sociology*, 24, 312–343.

Forbes M.H., Bielefeldt A.R., & Sullivan J.F. (2016). Implicit bias? Disparity in opportunities to select technical versus nontechnical courses in undergraduate engineering programs. ASEE Annual Conference and Exposition, Conference Proceedings, New Orleans, LA.

Holosko M.J., Barner J.R., & Allen J.L. (2016). Citation impact of women in social work: Exploring gender and research culture. *Research on Social Work Practice*, 26, 723–729.

Jones W.A. (2016). Do college rankings matter? Examining the influence of "America's best black colleges" on HBCU undergraduate admissions. *American Journal of Education*, 122, 247–265.

Leiby R. (2014). The U.S. News college rankings guru. *The Washington Post*, September 9. 2014, https://www.washingtonpost.com/lifestyle/style/the-us-news-college-rankings -guru/2014/09/09/318e3370-3856-11e4-8601-97ba88884ffd_story.html.

Meredith M. (2004). Why do universities compete in the ratings game? An empirical analysis of the effects of the U.S. News and World Report college rankings. *Research in Higher Education*, 45, 443–461.

Morse R., Krivian A., & Jackwin A. (2016). How U.S. News calculated the Best Global Universities rankings, https://www.usnews.com/education/best-global-universities /articles/methodology.

Porter T.M. (1995). *Trust in Numbers*. Princeton University Press, Princeton, NJ.

Redden E. (2014). "U.S. News" to issue new global university rankings, https://www.inside highered.com/quicktakes/2014/10/10/us-news-issue-new-global-university-rankings.

Sauder M. (2006). Third parties and status position: How the characteristics of status systems matter. *Theory and Society*, 35, 299–321.

Sauder M. & Lancaster R. (2006). Do rankings matter? The effects of *U.S. News & World Report* rankings on the admissions process of law schools. *Law and Society Review*, 40, 105–134.

Shehatta I. & Mahmood K. (2016). Correlation among top 100 universities in the major six global rankings: Policy implications. *Scientometrics*, 109, 1231–1254.

Smith S. (2013). U.S. News pulls social levers to break records for "Best Colleges" package, http://www.minonline.com/u-s-news-pulls-social-levers-to-break-records-for-best -colleges-package/#.U9fIXfldXTo.

Sumner D.E. (2012). American magazine winners and losers: 2001 to 2010. *International Conference on Communication, Media, Technology and Design (ICCMTD)*, May 9–11, 2012, Istanbul, Turkey, pp. 37–41, http://www.cmdconf.net/2012/makale/6.pdf.

THE (Times Higher Education) (2014). Times Higher Education announces reforms to its World University Rankings, https://www.timeshighereducation.com/world-university -rankings/news/times-higher-education-announces-reforms-to-world-university -rankings.

U.S. News & World Report (2013). Celebrating 80 Years: A timeline of events in the life of U.S. News & World Report, 1933–2013, http://www.usnews.com/info/articles /2013/05/17/celebrating-80-years.

Chapter 13

University Performance in the Age of Research Analytics

Francisco J. Cantú-Ortiz and James Fangmeyer Jr.

Tecnológico de Monterrey, Mexico

Contents

13.1 Introduction

Models take some traits, processes, or phenomena of a real entity and are used to study, predict, or modify the behavior of that entity. University rankings are models of higher education systems. Modeling is a common activity in Data Analytics, and university rankings are one well-known activity of Research Analytics. In this final chapter, we summarize lessons learned from the previous 12 chapters. These show us the Bibliometric databases on which Research Analytics is built. These also display how ranking models are built, expose the mechanisms within these competing models, and tell us how to boost performance and competiveness of education institutions with these models. It is said in Chapter 1 that Research Analytics is the connection of Data Analytics, Bibliometrics, Scientometrics, and research intelligence in both prospective and prescriptive ways. The contributors of this book have presented the state of the art in Research Analytics.

This chapter posits future trends for Bibliometric databases and rankings of world-class universities and then explores the strategies that universities can use to succeed in the age of Research Analytics. Finally, it considers ethics as foundational to higher education and especially to Research Analytics. Section 13.5 briefly describes the obligations of each stakeholder in the research analysis enterprise. The final conclusions offered in this chapter are but a checkpoint in the development of Research Analytics. In the meantime, they may provide ways to boost university competitiveness with Scientometrics.

13.2 Bibliometric Databases: Future Trends

This section briefly reviews Web of Science (WoS), Scopus, and Google Scholar (GS) and offers one perspective on where these key Bibliometric databases are heading. It also summarizes institutional repositories and speculates on the future of this trending technology. These four sources of Bibliometric data will continue to expand their scope of coverage and offering of indicators.

13.2.1 Major Bibliometric Players

The three major Bibliometric databases, WoS, Scopus, and GS, are establishing distinct competitive strengths. WoS is currently making the biggest move because it has changed ownership from Thomson Reuters to Clarivate Analytics. What was once a small part of a huge conglomerate has become the sole focus of a new and agile organization. Immediate changes have taken place in branding. Clarivate Analytics markets itself as an impulse to innovation, especially the discovery, protection, and commercialization of knowledge. This commercial focus could become the competitive advantage of WoS and create an interesting fusion of economic, technology, and research analytics coming out of this database. We should expect some upgrades to the database itself, which sports older technology than the newer Scopus database. The position of a private equity company in the ownership of Clarivate Analytics may suggest that the product will be turned around and sold again at a profit. Even with technological and business changes, WoS has maintained commitment to its *Emerging Sources Citation Index*, which in the future will provide Bibliometric data on a greater number of sources. WoS historically focused on natural sciences. The planned expansion will cover more social sciences and more sources not published in English.

Scopus has quickly grown into a Bibliometric database able to compete with WoS. As a result of this rise, Elsevier, the company that owns Scopus, has become a juggernaut in the research industry. The competitive advantage that Elsevier offers is an ecosystem of research products that includes Scopus, SciVal, Mendeley, ScienceDirect, and its collection of academic journals. There exists a moral hazard for Elsevier because it is both the publisher of academic journals and the owner of Scopus, which gives reputation and commercial boosts to the academic journals that it indexes. Overall, the entire ecosystem creates a lot of value for researchers, but some conflicts of interest must be navigated.

GS differentiates itself from Scopus and WoS by being a completely free and open service. GS also has the distinction of indexing the entire World Wide Web (WWW), and not just a collection of journals and conferences. GS is not likely to abandon either one of these defining product features. GS has not yet indicated that it intends to become a service or a solution provider, which is a movement that Clarivate Analytics and Elsevier are making.

A future trend for Clarivate Analytics and Elsevier is the sale of research services and solutions. Clarivate Analytics currently sells InCites, and Elsevier sells SciVal, to serve the analysis and intelligence needs of their clients, mainly universities and research institutes that need to evaluate competition as well as their own progress. These are software-as-a-service dashboards that query each company's Bibliometric database and return customized charts and graphs to the user. The next generation of services includes Clarivate's Converis and Elsevier's Pure. These are entire research information systems to be integrated with university IT architecture. The benefit of buying such a research information system is the integration of an institution's internal research database with one of the world's leading Bibliometric databases. This facilitates data capture, strengthens data integrity, and permits data comparison. Clarivate Analytics and Elsevier are beginning to sell these services because there is an increasing demand for Current Research Information Systems (CRIS).

CRIS have been under development at universities for at least 20 years. Their future trend is continued growth, either by university in-house development or via for-sale services from a Bibliometric database. These systems capture all research production at universities and make it public to the extent legally and practically possible. More universities will have such systems in the future, and such systems will contain full access to an ever greater number of university products. These products include not only peer-reviewed articles, but also theses, artwork, lectures, and more. CRIS require IT expertise that can be developed in-house or contracted from a Bibliometric database company. Another ongoing challenge is the proper respect for publisher rights and diffusion of full-text documents.

13.2.2 Digital Identifiers

Veracity, or truthfulness, is one of the V's of Big Data, and it is a trend in Bibliometric databases regarding data disambiguation. Name variants are a huge problem for researchers, institutions, and analysts. The root of the problem is that human language may refer to the same object with different words. Sometimes, a person may use two names, sometimes three, but the person does not change. Sometimes, an institution may use its full name and sometimes an abbreviation, but the institution does not change. Databases often do not understand this ambiguity. The repercussion is that true data about a professor or an institution become scattered under various name variants and are hard to unify again into a complete picture.

Several ID systems have been developed to solve this challenge. Major publication databases have their own internal ID systems. Scopus has Scopus Author ID, WoS has RID, and PubMed has PubMedID. These only serve individuals and do not transcend databases. A larger effort called Open Researcher and Contributor ID (ORCID) attempts to place a single, permanent, and transcendent ID number on researchers and research organizations whenever they operate digitally. ORCID is being integrated with more databases all the time.

In some ways, ORCID reflects a preexisting system called Digital Object Identifier (DOI). DOI is a collaborative effort by research stakeholders to put a unique ID code on digital objects, typically published research. This makes comparisons between publication databases easier. It is often the responsibility of the journal that publishes a document online to request a DOI for that document.

ORCID and DOI may clarify the identity of institutions, researchers, and publications, but the relationships between these three are still very hard to record correctly because researchers move between institutions. An author changing institutions should continue to carry a composite history of work done at both, but each individual publication must exclusively be affiliated to one institution or the other (assuming that the author does not maintain two affiliations). This is especially challenging when the author moves during the review and publishing process. Different databases will affiliate differently a document which carries a footnote explaining that an author now has an updated address.

It is in the interest of all stakeholders, especially the research analyst and the researcher, that one universal standard be established in the digital identifier challenge. This will serve to widen the scope of easily comparable data and simplify data analysis. Digital identifiers are an effort that includes engineers of the WWW, government agencies, and academics. A superior digital identifier developed to solve ambiguity problems in research could also have applications in other industries where keeping digital track of entities and their relationships is difficult.

13.2.3 Internet of Papers

The rise of artificial intelligence experienced in other industries will also come to permeate academia and research. Deep learning is a technology that allows computers to infer abstractions out of millions of small details (Lecun et al., 2015). This permits computers to perform tasks that may be described as conceptual rather than repetitive computational legwork. Computers trained with artificial intelligence are becoming so powerful that they are beginning to replace scientists. *Adam* and *Eve* are prototype robot scientists that have executed all aspects of the scientific discovery process (Sparkes et al., 2010). They generate hypotheses, plan adequate methodologies, perform experiments with robotic arms, analyze data, and interpret results (Buchen, 2009). Intelligent physical machines such as Adam and Eve will become more common. The Internet of Things is the name for the trend in which more physical objects have computer chips programmed to capture and respond to data in intelligent ways.

The *Internet of Papers* (IoP) is a trend to anticipate inside the rich digital bits of Bibliometric databases.* The IoP will be a software tool that works for, and eventually works with, the researcher. In the Internet of Things, objects in warehouses are attached with sensors and small processors to communicate with other objects in the warehouse and with the entire assembly line in which they participate. In the IoP, each digital object inside a Bibliometric database will have a program attached to it, and it will discuss its contents with other papers and with the database. Massive deep learning applied over the published knowledge inside a Bibliometric database will offer powerful services to researchers. A deep learning IoP application could find background literature on a subject and prevent plagiarism. As a result of learning hundreds of previous experiments, the deep learning IoP application could propose justifiable prior distributions for use in Bayesian statistical inference (Seaman et al., 2012). As it continues to learn from papers that are added to the database, it could connect relevant knowledge to a researcher's conclusions and even propose new lines of research. Artificial intelligence capabilities already exist such that the IoP application could write its own academic articles.

In 1900, it was difficult to imagine that machines would replace horses in terrestrial transportation. Today, it is difficult to imagine a machine replacing a human in research and development, but it is possible that machines will contribute to scientific progress. Moreover, it is probable that something like a deep learning IoP application will unlock the discoveries stored in today's Bibliometric databases.

---

* First presented in the paper by Cantú (2015). Elaborated in the paper by Cantú et al. (2016).

13.3 Rankings: Future Trends

Rankings are models of higher education reality. The major trends to watch that are affecting these models are world university rankings (WUR), regional rankings, national rankings, subject rankings, beyond-research rankings, alternative metrics, quantitative versus qualitative assessment, and research social networks. Each of these trends is briefly elaborated in this section to introduce the reader to its implications.

13.3.1 World University Rankings

The competition between ranking agencies has helped each to develop a distinguishing factor of its ranking. Competition between rankings is affected by readership as well as by academic investigation and commentary. Academic research may focus on methodological details, as in the study by Zhang et al. (2014), or on societal impact, as does much of the research reported in this book's chapter on *U.S. News & World Report* Best Global Universities (BGU) ranking. In response to consumer and academic feedback, ranking agencies are distinguishing themselves in order to remain competitive.

Quacquarelli Symonds (QS) is specializing in graduate employability. It includes employer reputation in its WUR, and it has a nonresearch-based Graduate Employability Ranking (GER). Times Higher Education (THE) is defining the university–industry relationship, as measured by industry income, as a distinguishing feature of its ranking. U.S. News brings a long specialization in the United States to its new international ranking, which for now is concentrated on Bibliometrics and has indicators that detect work in the absolute highest citation percentiles. Academic Ranking of World Universities (ARWU) has maintained its Scientometric-only construction since inception. Its special dimensions include *Nature* and *Science* publications and Nobel Prize winners. Now, it also offers simulated rankings in which it drops certain demanding criteria such as Nobel Prize winners. Scimago is pioneering an innovation-based ranking that includes patents granted and citation flows from patents. This approach allows Scimago to include more government agencies and companies in its Research Analytics. In its 2016 Institutional Ranking, the first place goes to Centre National de la Recherche Scientifique (France), followed by the Chinese Academy of Sciences (China), Harvard University (United States), National Institutes of Health (United States), and Google Inc. (United States) (Scimago Lab, 2016). Finally, U-Multirank (UMR) is pioneering a user-centric ranking product that allows each user to choose which criteria to include as well as the weights assigned to each criterion. As a side effect of not publishing standard results, UMR does not get media coverage and is therefore less famous than its peers.

Ranking agencies are competing for survival, often over unexploited niche markets. One example of an ongoing niche competition exists between QS and THE in their rankings of universities under 50 year old. Both agencies are committing growing amounts of resources to this niche of young universities. THE has expanded from its original 100 under 50 to 200 under 50 for 2017. QS now offers the Next 50 Under 50 to cover more young institutions. The agencies have different strategies in this niche. QS uses the same methodology from its WUR for its under 50 rankings, limiting the value of its under 50 list. THE also recycles indicators from its WUR but weighs teaching and research reputation less heavily in the under 50 rankings. This is one example demonstrating the ways that competition and selection shape ranking agencies.

13.3.2 Regional Rankings

Regional rankings are as much a future trend as they are a historical trend. Regional disparities exist between countries and institutions as a result of different awareness and competitive priorities (Hijazi et al., 2014). The 2006 Berlin Principles for higher education rankings specified that linguistic, cultural, economic, and historical contexts of different institutions should be recognized and respected by rankings. The principles generally exhort relevant and comparable data sources (Institute for Higher Education Policy, 2006). This counsel was continued by Shin & Toutkoushian (2011), who proposed that global ranking systems be made more regional in order to reflect different characteristics found across the higher education landscape. This is a methodological and marketing issue.

Methods of regional rankings should differ from national and global ranking methods. One methodological difference at the regional scale versus the national scale is the availability of comparable data. *U.S. News & World Report* has the benefit in its national ranking of standard data reported through accrediting and governing bodies that oversee all institutions in the ranking. A ranking attempting to compare schools regionally across the entire North America would not have this luxury. This limits the types of analysis that can be done at the regional level and often promotes the use of Bibliometrics, which are standard internationally.

Regional ranking methods should differ from global ranking methods in ways that capture specialized characteristics. One temptation is to create a regional ranking by taking the same methodology and applying it the same way across regions. This is simply a filter based on geographic coordinates. For example, the U.S. News BGU ranking has a regional ranking component, but it is based entirely on the global methodology. Criteria and weights are the same in each region, which does not provide the user any more information than the overall BGU compiled rank. Unique criteria are often appropriate in diverse regions. In some higher education systems, it is taken for granted that instructors imparting bachelor's level courses have a doctoral degree. This is not a certain assumption in Latin America, and QS includes *proportion of staff with PhD* as a special criterion in its Latin America ranking. It is appropriate to adjust weights as well to reflect regional education priorities. Compared to its WUR, QS weights *academic reputation* lower and *employer reputation* higher in its Latin American ranking. This change recognizes that few Latin American universities are committed to being global research universities and respects the reality in this region of the world.

Regional rankings are rising in popularity in response to market trends. First, users increasingly consider international study options. Regional rankings increase consciousness about nearby options. Second, regional competitiveness is a strong geo-political reality. Countries and higher education systems measure themselves within a set of peer nations. Third, greater scientific production and greater access to Bibliometric and other digital standard data make ranking across borders more feasible. Technology is opening the door to a wider perspective. Fourth, ranking agencies are following users and data providers into the regional scope. They are offering more benefit to consumers and reaping wider media and commercial gains.

13.3.3 National Rankings

A year-old trend led by THE is the formation of national rankings with local partner organizations. For example, THE released the first *Wall Street Journal—THE* U.S. university ranking in 2016. Times just announced its inaugural THE Japan University Ranking in partnership with Benesse, a Japanese education company (THE, 2017). This is an emerging trend with commercial inspirations and methodological Research Analytic implications.

As mentioned previously, national rankings offer a more standardized basket of data. Schools are often impelled to report data by the government or accrediting agency, and this facilitates analysis. Additionally, social and economic metrics such as salary value added, loan repayment rate, teacher/student interaction, and student diversity can be compared within the same national context. The U.S. and Japan rankings have four key areas: resources, engagement, outcomes, and diversity. These are weighted differently in each country to reflect what matters most to students and other stakeholders. Both of these rankings focus more on teaching than research. Countries with large and wealthy higher education systems may be the only countries fortunate enough to receive a tailor-made ranking from THE.

13.3.4 Subject Rankings

Subject rankings offer the most useful models of higher education reality to students and parents. As of March 2017, QS publishes rankings of 46 subjects, THE ranks 11 subjects, and U.S. News ranks graduate programs in 37 subject disciplines. These are the best models for students and parents because they reduce noise while maximizing signal; they inform students with a limited scope of information that is highly relevant. Subject rankings take advantage of user initiative. When the user seeks a defined subject area, the user reveals some of his or her preferences and can therefore be targeted with better results. Subject rankings are also favorable to universities. First, subject rankings align with traditional university structures. Second, subject rankings give universities agility.

Universities are collections of faculties divided into subject matters. In this sense, subject ranking information is easily extracted from the university and assimilated into the university. Subject rankings serve as evaluation and planning tools that fit into organizational structures traditionally in place at universities.

Tradition may seem antithetical to agility, but subject rankings also provide agility to universities. Moving up in the WUR is extremely difficult because indicators such as publication output require large financial investments, citations indicators take years of patience to accumulate, and reputation indicators are notoriously monolithic. At the subject level, universities have the agility to target strengths, take strategic action, and improve their rank. Actions at the departmental level are within the scope and capacity of several dedicated leaders. The agile university has the opportunity to optimize for subject rankings and shine on an international stage. This makes them especially useful.

Although useful, subject rankings may disappear before regional rankings. As research tends toward interdisciplinary challenges, it remains to be seen what will become of subject rankings. Subject rankings excellently model the traditional higher education organization, but they may not be able to follow where higher education is going in the future.

13.3.5 Beyond Research Rankings

Dedicating a section to rankings that venture beyond research analysis is a testament to how dominant the model of global research rankings has become. Rankings discussed here are departures from the model of global research university ranking. Examples include Webometrics, uniRank, and Green Metrics. Examples produced by the major global rankers include THE Global University Employability Rankings, QS Graduate Employability Ranking, QS Best Student Cities, and QS Stars Rating.

Webometrics uses Internet link data to assess a university's connection and influence via its digital presence in WWW. Its claim is that reputation surveys are biased and that Bibliometric data are limited, and that as WWW encompasses more facets of human life, a university's web presence more accurately represents that institution. UniRank is a close competitor to Webometrics that ranks higher education institutions by the success of their websites in terms of traffic, trust, and quality link popularity. UniRank does not distinguish between research-related web content and other types of university content. It does not even distinguish if the content is academic-related. All content connected to the university's root domain is counted.

Green Metrics compares institutions based on sustainability and environmental impact worldwide. The focus is completely tangential to teaching and research and can be connected to the *third mission* of universities for social impact.

THE Global University Employability Ranking is produced in partnership with Emerging, a French human resources consultancy. Emerging developed the survey instrument that is administered to a panel of manager-level recruiters as well as a panel of managing directors of international companies. Sampling is done considering business sector as well as each country's number of students, number of institutions, and GDP. Respondents are asked to opine on universities that produce the best employees.

QS Graduate Employability Ranking ranks universities based on how they positively contribute to the personal economy of their graduates as well as the economic climate of their community. The indicators in the ranking are (1) employer reputation as measured by a survey, (2) the count of job offers in the university's job bank, (3) alumni outcomes, which include the number of alumni mentioned in the media for business leadership, (4) the count of events on campus sponsored by companies, and (5) graduate employment rate (QS, 2016). The reality is that students, parents, and governments are seeing a college degree as an investment that should pay dividends in economic stability for the holder and economic growth for the society. The Graduate Employability Ranking eschews research metrics in favor of metrics aligned with the view of higher education as an investment in the economy.

The QS Best Student Cities ranking covers 75 cities around the world and judges them based on (1) number and rank of universities in the city, (2) student counts and ratios, (3) various city *liveability* and quality indices, (4) national and international employers' opinions in the QS employer survey, (5) various costs of education and cost of living metrics, and (6) a student experience survey (QS, 2017). This analysis, very interestingly, makes nearby universities allies in determining the ranking fate of their city. This is in contrast to most rankings, which pit each university against all others.

The QS Stars Rating awards universities up to five stars for excellence as compared to fixed benchmarks, not relative to the performance of other universities. These benchmarks go beyond research indicators into employability, facilities, community engagement, and other areas not normally covered in rankings. In contrast to a ranking, this rating system does not put universities directly in competition. Each university can receive five stars independently of the results of the others. This system functions like the star rating system of hotels, in which stars signify a high level of quality. There are many ways to achieve the points necessary for five stars, including combinations that do not rely on outstanding research scores.

Specialized rankings reach special audiences. Webometrics and uniRank galvanize university IT departments to simplify domain names and optimize digital architecture for search and sharing. Green Metrics speaks to socially conscious demographics as well as university physical plant managers. Best Student Cities gets the attention of local governments, city communities, and students looking to study abroad. Employability rankings speak to the investment minded. The Stars Rating creates a space where universities are recognized for excellence, but not at the expense of their peers. These rankings can continue to grow as long as these special audiences exist, and unserved audiences will become easier to serve as more data come online.

Nonresearch universities clamor for their important contributions to be recognized and honored. Of key interest is measuring university social and economic impact in local and regional communities. Beyond-research rankings will likely become more popular, but not at the expense of research rankings. These two types of rankings often interest the same audience, and each has the benefit of its separate niche following. There are sufficient data and readership in the higher education industry for both to grow.

13.3.6 Alternative Metrics

Alternative metrics may only be called such because a set of *traditional* metrics has become entrenched. Traditional metrics quantify research outputs such as publications, citations, journal impact factors, and coauthors. We consider alternative metrics as quantifiable attributes of research that are outside the set of traditional measures. Of course, an alternative metric that is tested, adopted, and popularized ceases to be alternative and joins its traditional predecessors. Alternative metrics are being proposed every year, and they are tested in Scientometric research and public platforms. One example is the breakthrough paper indicator, which was released in 2012 (Ponomarev et al., 2012) and was rereleased after further research in a 2.0 version in 2014 (Ponomarev et al., 2014).

Popular alternative metrics today include digital events, networks, and economic impact. Like so many industries, research is moving online. Digital events such as article views and downloads are tracked as easily as publications and citations. These digital events are considered to be close reads on the pulse of research. It is being studied whether views and downloads are predictors of later citations. Graph theory and social network analysis are applied to research today because of its increasingly connected and collaborative nature. Common networks to analyze are coauthorship and cocitation networks, which can help authors find collaborators, help laboratories identify trending topics, and help universities take strategic global positions. Social media networks such as Twitter are also becoming sources of alternative metrics. Research rarely reaches the general public, and therefore, the shares and likes of research-related posts signal special activity. Another way of seeing research's impact outside of the academy is through economic indicators such as patents. Engineering, science, and medical schools are increasingly interested in the number of patents generated by their faculty and the citations attributed to their research that come from newly granted patents in industry.

Alternative metrics are the frontier of Research Analytics. New data sources and new methods are subject to investigation that should conclude whether these extensions are useful measures of science. Thanks to the wide range of stakeholders in higher education, the range of alternative metrics is wide and changing. Considering alternative metrics today allows universities to extract competitive intelligence gives them advantages in the future.

13.3.7 *Quantitative versus Qualitative Assessment*

Both quantitative and qualitative assessment will be part of university rankings in the future. Generating quantitative metrics will be subsidized by ever-growing data capture and storage facilities. Interest in qualitative measures will persist as long as humans enjoy passing judgment on their environments.

Number crunching, statistics, and quantitative metrics are what come to mind when first contemplating Research Analytics, and the quantifiable is fundamental to Research Analytics. Like science itself, Research Analytics should be verifiable and replicable, and quantities help that. Nevertheless, many of research's quantifiable metrics have their origins in something more qualitative. For example, citations are verifiable and replicable quantitative data extracted from scientific databases, but they interest us because they are approximations of a qualitative judgment. Citations are considered a measure of quality. A citation is a quantitative binary outcome of the author's qualitative perception of the preexisting literature. If the quality is perceived highly enough, the result of the function is one citation. Even the existence of preexisting literature is the result of several quality perceptions made by reviewing editors of a journal. It is useful to remember that quantitative measures such as citations and publications have qualitative roots because it helps us understand the ways that these metrics may be biased.

Qualitative judgment will continue to interest stakeholders in the research enterprise. It is part of human nature to pass qualitative judgment and to value perception. Nevertheless, reputation surveys are maligned as biased. Perhaps, it is because of this tension that qualitative assessments in rankings have put on quantitative clothing. To be considered as legitimate evidence, perceptions must be collected by validated survey instruments and treated with proven statistical methods. Rankers are happy to oblige. Rankers are promoters of reputation surveys because these have become part of their commercial competitive advantage. Increasingly, consumers with Internet access can run a query and see for themselves which of two universities has more publications, but the same consumer does not have the resources to conduct a reputation survey. Rankers have built apparatuses to ask thousands of experts for their opinions and are storing the results in private files. For this reason, it does not seem that rankers will let go of qualitative surveys. Neither is the public entirely interested in doing so. Thus, qualitative assessment will be a part of Research Analytics along with quantitative assessment in the future.

13.3.8 *Research Social Networks*

Research social networks are worth considering by research analysts as sources of data. They also serve as sources of data for researchers. Research social networks that do not serve as good data providers to either of these audiences have failed.

ResearchGate is the largest network of its kind. It has an interface similar to Facebook or LinkedIn, but a target audience limited to research stakeholders that can publish profiles with professional information. Research analysts scrape this information for numbers of views, numbers of downloads, other Bibliometric indicators, and the proprietary *ResearchGate Score* of a researcher's work (Orduña-Malea et al., 2016). An analyst may also see connections, following coauthor relationships on ResearchGate. Fellow scientists may use the network to share datasets, request full-text articles, or keep in touch.

GS placed an automatic approach on the construction of a research social network. Readers may object that GS is not a social network. It certainly lacks the look and feel of ResearchGate, but it is built on the largest network in the world, the WWW, and uses profiles to calculate coauthorship connections. Furthermore, it is a vast source of data for research analysts and researchers. GS automatically reports the citations and *h*-index for its profiles from data across the entire WWW. Analysts at Webometrics have used the data available from GS to rank individual researchers (Cybermetrics Lab–CSIC, 2017). Researchers get data including institutional affiliation, publication metadata, and citation counts from GS.

Some social networks such as Microsoft Academic and Academia.edu have failed in part because they could not provide enough useful data to relevant audiences. Mendeley is another platform that aspires to be a useful social network and must consider serving both research analysts and researchers going forward. No major global university ranking explicitly uses research social networks as a data source, but these networks have the potential to redistribute power among research stakeholders and are a trend to watch.

13.4 University Strategies in the Age of Research Analytics

Universities are changing their interinstitutional and intrainstitutional behavior thanks to Research Analytics. Intrainstitutional changes are those that do not affect other institutions, but may affect individual stakeholders of the university such as professors and students. Interinstitutional changes are those that directly involve other institutions such as other universities, media, and ranking agencies. Academic research is evolving into a more shared and networked enterprise, and Research Analytics is helping universities adapt to this phenomenon (Aguillo, 2015). Research offices are a manifestation of the new relevance of Research Analytics in university strategy. These offices often report to the provost and transcend any faculty or school. They collect and analyze research and academic information within universities to supply ranking agencies with input data which includes number of students, professors, research funding, and revenue. These administrative functions are a layer removed from the actual research at a university, but attempt to create strategies that strengthen this pillar of the institution as well as document and evaluate progress.

13.4.1 Intrainstitutional Strategies

A proliferating strategy in the age of Research Analytics is to use analysis, especially published rankings, to define and measure key process indicators (KPIs), and to identify the actions necessary to improve these KPIs. A cynic may argue that this strategy is a capitulation by the university to the rankings. Building a university around ranking criteria is like *teaching to the test*, they may say. An optimist, or perhaps a realist, would argue that universities are intrinsically motivated to improve research, teaching, and social impact and that adopting this strategy simply leverages a strong extrinsic motivation to achieve the same goals. Analyses such as rankings help universities track and compare elements of research across time and across peer institutions. This does, in fact, give more information to researchers, deans, and trustees. These actors have reacted to ranking metrics with internal strategies including the examples in the following.

The ratio of students/faculty that is part of several rankings has put strategic pressure on universities. This metric is supposed to approximate the quality of teaching at a university. Using this metric as a strategic KPI may or may not achieve the goal of increasing teaching quality. Hiring and admission strategies have been directly affected by the student/faculty ratio. One pressure of this metric is for the university to hire more faculty, regardless of how well these faculty teach. The result could be that a university ends up with more faculty, stretched thin for resources, who do not know how to teach, and the quality of teaching actually drops. A second pressure of this metric is to admit fewer students, to a certain extent regardless of their ability to learn. The result could be the elimination of a subset of the student population, and this will have ripple and unintended consequences across the institution. Research output is indirectly affected because students and faculty are two pools of human resources that produce the research at a university. Changing the ratio of students and faculty impacts internal research structures and their overall effectiveness.

Other ratios in rankings such as citations/paper and citations/faculty are directly related to research strategies and are therefore very powerful, for better or for worse, when used as KPIs. Citations, considering some normalization factors by discipline, are considered the mark of quality research. It is a goal in every university to increase citation counts, but each university must make the strategic decision *at what cost?* Two major costs in producing citations are papers and faculty.

The citations/paper ratio pressures universities to actually publish less, especially to publish less in low quartile journals or conferences. Research directors are cautioning their faculty to think cautiously about publication outlets and only to publish where they can expect high visibility, prestige, and citations. This begins to affect other stakeholders such as low quartile journals and certain conferences. Low quartile journals traditionally offered the value of a publication outlet for new researchers or for an incremental discovery. It is not uncommon that a work of this class has later become foundational for the science that is published in top quartile journals. Based on maximizing citations/paper, new researchers are pressured to publish in collaboration with an established name in a top journal. Incremental science, replications of experiments, and perhaps even failed experiments will not be published in the lower quartile journals that accepted this type of work. Conferences have always offered benefits other than citations to researchers. These include expanding cultural horizons, networking with colleagues, and receiving feedback on incomplete projects. A researcher or university too focused on citations/publication will lose these rich benefits offered by conferences.

Using citations/faculty as a KPI puts humans on the line. On the positive side, it pressures university administrators to invest significantly in the faculty that they have. Aforementioned research offices are busy mining data for ways to maximize the impact of each researcher. Diverse interinstitutional strategies include funding laboratories based on projected productivity, setting research priorities based on keyword analysis, and encouraging collaboration based on coauthorship networks. Many faculty will notice the revived interest that administration has in their work and productivity. The dark side of this KPI could include a hiring freeze, personnel cuts, or a redefinition of the faculty. The publish or perish paradigm may become *be cited or perish.*

Academic reputation is a KPI of paramount importance to universities. One strategy that has arisen to improve academic reputation is greater emphasis on English even in countries where English is not the first language. The reason is that English is the dominant language of science and technology, and rankings are weighted toward scientific and technological production. The current scientific system is biased to give visibility to work in English. One internal strategic response, therefore, is to hire faculty with English capabilities and to attract and train students with English capabilities. To do this, universities across the globe offer degree programs in English for students as well as English language classes for researchers. Another internal response is to translate university web pages into English. Divulgation must be done in English in order to persuade the wealthiest and most influential academic audiences. Universities outside of English-speaking countries are at a relative disadvantage in the KPI of academic reputation and must make strategic compensation.

Universities have a good idea about their performance even without the metric barrage of rankings, and many universities are exceptionally valuable in ways that rankings do not capture. History and current context make elite universities in each country different (Kaiser & Vossensteyn, 2012). Research Analytics including rankings can offer another tool to help each university pursue progress against its own independently defined mission. All stakeholders are shamed if a university loses its cultural and contextual character in an effort to game the global research university rankings. If a university is sure in its objectives, then using Research Analytics in its KPIs can be a helpful way to build strengths and cure weaknesses.

13.4.2 *Interinstitutional Strategies*

Higher education institutions maintain relationships with each other and with intermediary institutions such as media and ranking agencies, and Research Analytics is changing the strategies that guide these relationships. As these are relationships between two institutions, they affect two large populations of individuals who participate in the institutions. The decisions that impact all of the participants are concentrated in the hands of interinstitutional brokers, who have a great responsibility to craft strategies for positive change.

The first issue is academic reputation. Academic reputation is typically measured by a peer survey such that each institution depends on others for its reputation. This has propelled universities to invest in their brands, especially as perceived by other universities. A principle tactic is signing formal collaboration agreements, not only on research or in dual-degree programs, but also on faculty and even student exchange. Importantly, formal connections link the reputations of two institutions and give the incentive to speak well of the partner institution both in reputation surveys and in professional circles. A more prestigious reputation is good for university stakeholders including students, faculty, alumni, employees, the city, and the country. The impact of collaboration can be visualized* as well as studied with statistical methods such as Bayesian multilevel logistic regression (Bornmann et al., 2016). Universities are using Research Analytics to identify peer communities in which they have the most influence and from which they can draw the greatest boost to their public reputation, and then they are setting interinstitutional strategies to improve their own position.

Media and ranking agencies are institutions that determine much of a university's brand and are therefore critical to a university's interinstitutional strategy. A university oblivious to media and rankings is allowing a large component of its own success to be determined without a strategy. The power of media and rankings should not be ignored, even if these institutions are unpleasant to university stakeholders. It is better for universities to enact a strategy than to leave their own reputation in the hands of others. The reputation created in media and rankings is often self-fulfilling, and therefore, an effective interinstitutional strategy for dealing with media and rankings can be an asset. A major component of a media and ranking strategy is a competent communication team. This team is competent in media from social networks to academic journals and uses each channel to deliver a positive image of the university. The team understands which audience is on the other end of each channel and shows its relevant information.

The world of Research Analytics is bigger than just universities. Universities are adapting to this world with interinstitutional strategies. They are beginning to take control of their own destinies with initiatives such as the Snowball Metric Project, which defines agreed-upon metrics and methods for Research Analytics

---

* Excellence Networks, http://www.excellence-networks.net/

(Colledge et al., 2012). This brief section has outlined just some of the strategic responses inside the academy and between institutions. More exciting new strategies are yet to come.

13.5 Ethics

Research has become a societal effort; each stakeholder cooperating in the effort is vulnerable to the misbehavior of stakeholders, which is why ethics is central to this book. The decision-making structure, in which each stakeholder is responsible for policing its own decisions and in which these decisions impact the entire society, makes ethics indispensable.

Bibliometric databases have become critical in the assessment of universities via Research Analytics. It is an ethical responsibility of these databases to remain impartial toward any school. It is also a commercial responsibility of these databases to remain impartial to all publishers. Scopus in particular will be scrutinized because it is owned by the same company, Elsevier, that publishes many academic journals. Journals published by Elsevier should not have preferential treatment when being evaluated by Scopus. CRIS should be lauded for their momentum toward open information and open science, but they must comply with the legal framework in place. These systems have an ethical responsibility to respect proper publishing and author rights.

The leaders of any university have an ethical responsibility to provide a certain level of transparency. Taken for granted is that universities should not report false data. Unfortunately, with increasing global pressure in the sector, educational institutions may be tempted to round numbers upward, define reporting periods and KPIs in ways that favor themselves, and even fabricate data. Another temptation for a university seeking global exposure may be to bribe a ranking agency. Universities offering bribes for better rankings should either be denounced publically or be excluded from the ranking, whichever is more likely to correct this unethical behavior.

Governments must respond ethically to rankings regarding the funds that they provide to universities. This theme is of greater or lesser importance depending on each country's higher education model. In some countries, the most productive research institutions are private, and in other countries, these institutions are public. In almost all research institutions, some money is received from a federal, state, or local government. These governments must refrain from taking politically motivated actions under the guise of sincere performance reviews tied to university ranking results.

Ranking agencies have ethical responsibilities to universities, governments, students, and parents. Rankers should actively avoid any situation that could compromise the objective nature of their results. Taken for granted is that rankers will not solicit or honor bribes. It is legitimate that a ranker have commercial interests. Rankers are within their rights to pursue profit. A profit motive, however, must never corrupt a ranker to compromise the objectivity of its ranking results. Rankers especially must not illicitly pressure universities that expect to be fairly judged regardless of commercial relationships with the ranker. Ranking agencies must separate completely their analytic operations from their business operations. Universities buying ads in ranking publications or hosting regional conferences in partnership with ranking agencies should receive no benefit or punishment in the ranking results. Rankers found breaking these rules should be boycotted by universities, denounced by governments, and ignored by readers.

Documentation led by the United Nations Educational, Scientific, and Cultural Organization has already related some uses and misuses of rankings in higher education (Marope et al., 2013). The opportunity is taken here to affirm previous work on ethics in research and invite yet more discussion in the future. Obviously, this discussion goes beyond rankings and includes Bibliometric databases, governments, university behavior, and responsible consumption by the public. Research is a societal investment, and society cannot permit that it be corrupted. If databases, universities, governments, and rankers play their parts, this will not be a problem.

13.6 Conclusion

This book presents current issues and future trends in Data Analytics applied to scientific research and the importance of this for world-class universities. As defined in this book, Research Analytics is the confluence of Data Analytics, Bibliometrics, Scientometrics, and research intelligence in descriptive, prospective, or prescriptive configurations. Research analytics is still an emerging field with the excitement of new breakthroughs, consolidation, and propagation. It is an important topic applied to strategy at world-class universities.

One focus of this book has been research databases. These impressive efforts advance science around the world. The databases Scopus, WoS, and GS were presented. Each of these is expanding to cover ever more scientific production. Among the biggest challenges that they face in the next decade are information disambiguation and the spread of open data. Artificial intelligence, especially deep learning, creates the world-changing opportunity to get new hypothesis and even new discoveries from the IoP.

A particular focus of this book has been the balanced use of rankings, promoting the positive aspects of rankings and their applications and warning against negative aspects and deviations. Following the wrong ranking, a university may wander far from its institutional mission. Following the right ranking, a university may find meaningful KPIs and develop useful strategies. Rankings presented include QS, THE, U.S. News, ARWU, Scimago, and UMR. In the current explosion of university rankings, this book is careful to consider today's leading rankings. This book did not cover rankings such as University Ranking by Academic Performance from Turkey, the NTU ranking, the CWTS Leiden ranking, and the Round University Ranking from Russia. Other rankings that look beyond research are only mentioned peripherally in this book. The relative importance of rankings is fluid, and one can expect that in a decade, the major and minor players will look different.

World-class universities not only are known for their research, but also have missions of teaching excellence and positive engagement with society. This volume ignores education analytics and social responsibility analytics mainly because number crunching in education and social responsibility is still difficult to achieve. The main reason that university rankings largely ignore these themes is the lack of reliable data. This is simply a current technical limitation that should be solved quickly, especially with data from MOOCs. *Or should it?* Higher education stakeholders can anticipate the rise of education and social responsibility analytics and can take lessons from the history of Research Analytics. There will be controversy, there may be abuses, and there will be meaningful and useful ways to analyze more parts of the university. This richer analytic ecosystem, in turn, will put new strategic pressures on institutes of higher education.

For now, research is the driving force of world-class universities. Research analytics will continue to grow and become common practice in university administration because it increases organizational competitiveness. Contributions to the field may come from Data Analytics, Bibliometrics, Scientometrics, or research intelligence. The greatest promise of a robust and ethical research analytic environment is its return to society, which today increasingly demands that higher education be made a public good. It is for the benefit of the planet, our countries, and those of us who have our lives transformed by research to bring Research Analytics into the future.

References

Aguillo, I. (2015). *Metrics 2.0 for Science 2.0.* Paper presented at 15th International Society of Scientometrics and Informetrics Conference (June 29, 2015, Istanbul, Turkey): Boğaziçi Üniversitesi.

Bornmann, L., Stefaner, M., de Moya Anegón, F., Mutz, R. (2016). Excellence networks in science: A web-based application based on Bayesian multilevel logistic regression (BMLR) for the identification of institutions collaborating successfully. *Journal of Informetrics* 10, pp. 312–327.

Buchen, L. (2009). Robot makes scientific discovery all by itself, *Wired*, April 2, 2009. Accessed at https://www.wired.com/2009/04/robotscientist/.

Cantu, F. (2015). Framing attributes and learning new patterns of collaboration from unstructured metadata in scientific publications. Keynote at the Data Analytics Summit II in Harrisburg, PA, December, 2015.

Cantu, F., Fangmeyer Jr., J., Ceballos, H. (2016). *The Internet of Papers.* Technical Report RIR-D-2016-08, Tecnológico de Monterrey, México.

Colledge, L., Clements, A., el Aisati, M., Rutherford, S. (2012). Project snowball—Sharing data for cross-institutional benchmarking. In Jeffery, K., & Dvořák, J. (Eds.) *E-Infrastructures for Research and Innovation: Linking Information Systems to Improve Scientific Knowledge Production*. Proceedings of the 11th International Conference on Current Research Information Systems (June 6–9, 2012, Prague). pp. 253–260. Accessed at http://www.eurocris.org.

Cybermetrics Lab–CSIC (Consejo Superior de Investigaciones Científicas) (2017). *Ranking of Researchers by Country (58 Countries)*. Accessed at http://www.webometrics.info /en/node/116.

Hijazi, S., Sowter, B., Nag, H. (2014). Higher education in MENA through global lenses: Lessons learned from international rankings. In Baporikar, N. (Ed.) *Handbook of Research on Higher Education in the MENA Region: Policy and Practice* (pp. 31–51). Hershey, PA: Information Science Reference.

Institute for Higher Education Policy. (2006). *Berlin Principles on Ranking of Higher Education Institutions*. UNESCO CEPES. Accessed on February 14, 2017, at http:// www.che.de/downloads/Berlin_Principles_IREG_534.pdf.

Kaiser, F., & Vossensteyn, H. (2012). Excellence in Dutch Higher Education: Handle with care. In Palfreyman, D., & Tapper, T. (Eds.) *Structuring Mass Higher Education: The Role of Elite Institutions* (pp. 169–181). New York: Routledge.

Lecun, Y., Bengio, Y., Hinton, G. (2015). Deep learning. *Nature* 521, pp. 436–444.

Marope, P.T.M., Wells, P.J., Hazelkorn, E. (Eds.). (2013). *Rankings and Accountability in Higher Education Uses and Misuses*. Paris: UNESCO. Accessed on Feb. 28, 2017, at http://unesdoc.unesco.org/images/0022/002207/220789e.pdf.

Orduña-Malea, E., Martín-Martín, A., Delgado-López-Cózar, E. (2016). ResearchGate como fuente de evaluación científica: Desvelando sus aplicaciones bibliométricas. *El profesional de la información* 25, pp. 303–310.

Ponomarev, I.V., Lawton, B.K., Williams, D.E., Schnell, J. (2014). Breakthrough paper indicator 2.0: Can geographical diversity and interdisciplinarity improve the accuracy of outstanding papers prediction? *Scientometrics* 100, pp. 755–765.

Ponomarev, I., Williams, D., Lawton, B., Cross, D.H., Seger, Y., Schnell, J., Haak, L. (2012). Breakthrough Paper Indicator: Early detection and measurement of groundbreaking research. In Jeffery, K., & Dvořák, J. (Eds.) *E-Infrastructures for Research and Innovation: Linking Information Systems to Improve Scientific Knowledge Production*. Proceedings of the 11th International Conference on Current Research Information Systems (June 6–9, 2012, Prague). pp. 295–304. Accessed at http://www.eurocris .org.

QS (Quacquarelli Symonds) (2016). *QS Graduate Employability Rankings Methodology*. Accessed at https://www.topuniversities.com/employability-rankings/methodology.

QS (2017). *QS Best Student Cities Methodology*. Accessed at https://www.topuniversities .com/best-student-cities/methodology.

Scimago Lab (2016). *Scimago Institutions Rankings*. Accessed at http://www.scimagoir.com /rankings.php.

Seaman, J.W., Seaman Jr., J.W., Stamey, J.D. (2012). Hidden dangers of specifying non-informative priors. *American Statistician* 66, pp. 77–84.

Shin, J.C., & Toutkoushian, R.K. (2011). The past, present, and future of university rankings. In Shin, J.C. et al. (Eds.) *University Rankings, The Changing Academy* (pp. 1–16). New York: Springer.

Sparkes, A., Aubrey, W., Byrne, E., Clare, A., Khan, M.N., Liakata, M., Markham, M. et al. (2010). Towards Robot Scientists for autonomous scientific discovery. *Automated Experimentation*, 2, p. 1.

THE (Times Higher Education) (2017). *Japan University Rankings 2017.* Accessed at https://www.timeshighereducation.com/world-university-rankings/launch-new -japan-university-ranking.

Zhang, Z., Cheng, Y., Liu, N.C. (2014). Comparison of the effect of mean-based method and z-score for field normalization of citations at the level of Web of Science subject categories. *Scientometrics* 101, pp. 1679–1693.

Index